Indirect Economic Impacts of Dams

THIS WORK IS PUBLISHED BY
ACADEMIC FOUNDATION, NEW DELHI
IN ARRANGEMENT WITH
THE INTERNATIONAL BANK FOR RECONSTRUCTION
AND DEVELOPMENT / THE WORLD BANK GROUP

Indirect Economic Impacts of Dams

Case Studies from India, Egypt and Brazil

Editors

RAMESH BHATIA
RITA CESTTI
MONICA SCATASTA
R.P.S. MALIK

THE WORLD BANK

ACADEMIC FOUNDATION

DISCLAIMER:

The findings, interpretations, and conclusions expressed in this volume do not necessarily reflect the views of the Executive Directors of The World Bank or the governments they represent. The World Bank does not guarantee the accuracy of the data included in this work. The boundaries, colors, denominations, and other information shown on any map in this work do not imply any judgement on the part of The World Bank concerning the legal status of any territory or the endorsement or acceptance of such boundaries.

Published in 2008
by
ACADEMIC FOUNDATION
4772-73 / 23 Bharat Ram Road, (23 Ansari Road),
Darya Ganj, New Delhi - 110 002 (India).
Telephone: 23245001 / 02 / 03 / 04.
Fax: +91-11-23245005.
E-mail: academic@vsnl.com
Internet: www.academicfoundation.com

Cataloging in Publication Data--DK
Courtesy: D.K. Agencies (P) Ltd. <docinfo@dkagencies.com>

Indirect economic impacts of dams : case studies from India, Egypt and Brazil / editors, Ramesh Bhatia ... [et al.].
p. cm.
Contributed articles.
Includes index.

ISBN 13: 9788171886814
ISBN 10: 8171886817

1. Dams--Economic aspects--India. 2. Dams--Economic aspects--Egypt. 3. Dams--Economic aspects--Brazil. I. Bhatia, Ramesh.

DDC 631.280 954 22

Printed and bound in India.

Contents

List of Tables, Figures, Boxes and Annexes

Tables

Figures

Boxes

Annexes

Editors/Contributors

Editors:

Ramesh Bhatia, until his untimely demise in February 2007, was President, Resources and Environment Group, New Delhi. He had long experience of working on international socio-economic and policy issues in the energy and water sectors. His past assignments included Energy Economist and Water Resources Specialist at the World Bank, Global Program Leader at the International Water Management Institute, Member and subsequently Senior Advisor to the Technical Advisory Committee of the Global Water Partnership, Professor of Economics at the Institute of Economic Growth, University of Delhi, India and a Visiting Professor at Harvard University. He also consulted with many international agencies, including the UNDP, UNICEF, ILO, and Asia-Pacific Development Centre.

Rita E. Cestti, is currently a Senior Rural Development Specialist in The Sustainable Development Department for the Latin America and the Caribbean region at the World Bank, Washington, D.C. She has managed the identification, preparation and supervision of a number of water-related, natural resources, environmental and disaster management projects in several countries and has led the preparation of several pieces of economic sector work. She has conducted extensive economic analysis in terms of sector work and project analysis as well as in-depth research in the areas of economics of water resources, demand management, water allocation, water pricing, water pollution control, and integrated planning.

R.P.S. Malik, is an economist currently working as a senior faculty member with the Agricultural Economics Research Centre,

University of Delhi, Delhi. He has more than three decades of research experience of working on various issues in agricultural and resource economics, notably on water resources, and has published extensively, including three books, in these areas. He has earlier worked for the World Bank, World Resources Institute, Afro- Asian Rural Development Organization and Agro Economic Research Centre, Shimla.

Monica Scatasta, is currently the Coordinator of OECD's Water Project on "Sustainable Financing to Ensure Affordable Access to Water and Sanitation". She is seconded to OECD from the European Investment Bank, where she is a Senior Economist in the Water and Environmental Protection Division. Previously, she has worked for the World Bank and IFPRI in Washington, D.C. She also worked in Brazil as part of the team supporting the Environment Ministry, National Water Agency and Paraíba do Sul River Basin Committee in the creation of the basin's water management system.

Contributors:

Meera Bhatia, is a doctoral candidate in Agriculture and Resource Economics at University of California at Davis. She has an M.S. in Agricultural and Resource Economics from University of Cailfornia at Davis and a B.S. in Environmental Economics and Policy from University of California at Berkeley.

John Briscoe, is currently World Bank Country Director for Brazil. Previously he was Senior Water Advisor of the World Bank with ressponsibility for the Bank's portfolio both globally and in South Asia. He holds a Ph.D. in Environmental Engineering from Harvard University.

Moataz El-Said, is with the Fiscal Affairs Department of the International Monetary Fund. He was ealier with the Trade and Macro Economic Division of the International Food Policy Research Institute (IFPRI), Washington, D.C. and with Cairo University.

Hans Lofgren, is with the Development Prospects Group in the World Bank since 2004. He was earlier a Senior Research Fellow at the International Food Policy Research Institute (IFPRI), Washington, D.C. He hold a Ph.D. in Economics from the University of Texas at Austin.

Alessandro Palmieri, is Lead Dam Specialist in the Corporate Quality Assurance and Compliance Unit of the World Bank. He advises the Bank on technical aspects of investment lending in the water sector, and provides quality management of the Bank Portfolio of Dam-related operations. He holds a Masters degree in Civil Engineering from the University of Rome—La Sapienza.

Sherman Robinson, is Professor of Economics at the University of Sussex, United Kingdom. He was earlier Director of the Trade and Macroeconomics Division and later Fellow at the International Food Policy Research Institute (IFPRI), Washington, D.C.

Kenneth M. Strzepek, is Professor in the Water Resources Engineering Department at the University of Colorado at Boulder. He holds a Ph.D. from Massachusetts Institute of Technology.

Preface

Questions on the indirect economic impacts of large dams lie at the heart of the development debate. Indirect benefits were the motivating factor behind most major infrastructure developments in the rich countries of the world. Rich countries now all have very large stocks of this infrastructure, which has provided the platform for economic development in these countries.

In a curious juxtaposition of narrow economic thinking and political convenience, for the last 40 years there has been a dominant view in the development community that (a) indirect economic impacts are impossible to measure and (b) the use of shadow prices (necessary because of hitherto large distortions in markets) took into account indirect effects.

This book takes stock of recent decades of development and development practice. It casts serious doubts on the claim that standard (shadow pricing-based) analysis as practiced does account for indirect benefits. And it shows that it is now possible, with the vast increase in computing power and the availability of calibrated general equilibrium and input-output models of the economy, to measure indirect impacts at reasonable cost and with considerable confidence. It also shows that as economic distortions in developing countries have been reduced, there is no longer a need for the extensive use of shadow prices.

These developments mean that economic analysis can return to being a transparent tool whereby policy makers and others can assess the indirect effects which are at the heart of the development process.

The book presents detailed descriptions of analyses of the indirect economic impacts of three major dams (Bhakra in India ; Aswan in Egypt, and Sobradinho in Brazil) and of a small dam (Bunga, in India). The general conclusions from the analysis are:

- That methodologies are now available which enable economists to estimate, ex post, the indirect effects of major water infrastructure projects and the incidence of these benefits on different groups.
- Indirect impacts are real and large, in the development stages in rich countries (Grand Coulee in the United States) and in contemporary developing countries (the studies of India, Egypt and Brazil in this book, and Malaysia);
- The incidence of benefits is quite different when indirect effects are added to direct effects, with some striking cases (Bhakra in India) where it is the poor, who were not beneficiaries of the direct impacts, who were the major beneficiaries—through the labour market—of the indirect impacts;
- The contribution to total (direct and indirect) benefits is often quite different ex-post than was predicted ex-ante. The most striking case in this book is that of the Aswan Dam in Egypt. The driving force behind construction of the dam was to develop Egyptian agriculture. But by far the largest benefits have emanated from electricity generation. There are two corollaries. First that it is very difficult, ex-ante, to predict the benefits of long-lived assets. Second, in such circumstances robustness is key—and multi-purpose dams are very robust producers of major streams of benefits as economies develop, as circumstances change and as societal values evolve.
- While the number of cases is small, the evidence in this study is that smaller dams (Bunga) indeed have smaller ripple effects and thus smaller indirect effects than large dams (Bhakra)
- While the literature is also small, there is evidence that the multipliers for infrastructural investments are much larger than for the "social" investments which are currently given highest priority by development agencies.

The work reported in this book has been carried out in stages over a period of four years and in the intervening period some changes in the underlying conditions prevailing at the four case study sites may have taken place. However, these changes are unlikely to make any significant difference in sending across the main message of the book: indirect economic impacts are significant and must be accounted for while quantifying the benefits of dam projects.

This book is completed with a heavy heart since Ramesh Bhatia, the first amongst equal authors of this book, passed away as the manuscript was being finalized. Ramesh was a brilliant economist and systems thinker who was equally adept at debating in the halls of the world's great universities, and in discussing with illiterate villagers why energy and water infrastructure mattered to them. He embodied that quintessentially Indian quality of being fully engaged, simultaneously and seamlessly, in the intellectual, ethical and spiritual domains. He was a mentor, friend and colleague to all of the other authors of this book, in some cases for thirty years, in others for a decade. As we completed the manuscript we kept wanting to pass revisions to Ramesh, so that we could be sure we got it right. We miss him, very deeply, and dedicate this book to him.

Last but not the least, we would like to thank the publishers Academic Foundation, New Delhi for their cooperation and help at various stages of the publication process of this book.

Rita Cestti, Monica Scatasta,
R.P.S. Malik, John Briscoe,
Alessandro Palmeiri

Part I

Methodological Issues and Overview of Results

1 Introduction

Ramesh Bhatia, R.P.S. Malik,
Rita Cestti and Monica Scatasta

Dams have brought considerable benefits to many countries and regions. They have enabled improvement and/or expansion of hydropower generation; irrigated agriculture; water supply for domestic and industrial uses; water for ecological services and pollution control; flood control; and reduced vulnerability to droughts. Dams have also provided active storage for the management of water, allowing countries and regions to strike a balance between natural river flows regimes and the pattern of water demand. In areas of the world where approximately two billion people live and where river flow is intermittent, flows are virtually unusable for development without storage. Large and growing populations in such areas require investments in large-scale surface storage. Of late, however, it has been widely recognised that there are substantial social and environmental costs associated with dams. As a result, the traditional view that dams have been an important source of social and economic growth has been challenged.

Dam assessment is by its nature a complex undertaking. First, many of the benefits and costs associated with dam development have quite different time streams, with hydropower and irrigated agriculture benefits materialising relatively quickly, flood control and drought reduction benefits being episodic, and some environmental costs only appearing decades later. Second, these benefits and costs are faced by different sectors and there are inter relationships between sectors so that, for example, agricultural expansion may bring in population that makes the region more resilient and able to take advantage of opportunities in industrial sectors at later times. Lastly, the effects of dams are distributed across different spatial scales, from local to basin, to regional to national, and in some cases, to transnational.

Table 1.1

Direct, Indirect and Induced Impacts of Dams

Outputs	
Flow regulation/flood control/drought prevention; hydropower generation, irrigation, water supply, fishing, recreation, tourism	Inundation of area, resettlement, change in hydrological regime of river, interruption of flow, reduction of navigable length
Benefits/Positive Impacts **Direct**	**Costs/Negative Impacts** **Direct**
1. Net benefits (¹) from direct—intended or unintended—project outputs: hydropower, irrigated agriculture, water supply, fishing, flood control (²), drought prevention, value of recreational activities/ tourism revenues. 2. Non-irrigation benefits of irrigation canals in terms of water for households, livestock, household enterprises, animals, birds. 3. Recharging of groundwater aquifers by canals. 4. Eco-system benefits, pollution control benefits and navigation benefits of flow control. 5. Economic and health benefits of reduced sediment/pollutant concentration in downstream reaches. 6. Other positive health impacts. 7. Reduced downstream pollution (e.g., sediment trapped in reservoir). 8. GHG* benefits of hydro power.	1. Social, economic, and cultural costs of resettlement. 2. Opportunity cost of lost output from inundated area. 3. Value of lost ecosystem/historical/ cultural heritage sites. 4. Losses from reduced numbers of fishes that spawn upstream. 5. Losses due to change in fishing conditions and technology. 6. Costs of introduction of foreign fish/ plant species. 7. Reduced nutrients in downstream river water when downstream users relied on them. 8. Reduced water availability or inappropriate timing of releases for downstream users. 9. Downstream pollution from point and non-point source pollution (irrigation, water supply users). 10. Navigation losses. 11. Negative health impacts. 12. Loss of GHG sink due to deforestation. Reservoirs as GHG sources? 13. Reduced efficiency if provide water below its long-run marginal cost, when that is the case. 14. For hydropower—cost and externalities of transmission, if expansion is needed to market project output.
Indirect (or Secondary)	**Indirect (or Secondary)**
1. Benefits (regional/inter-regional/national) from inter-sectoral production linkages: supply of inputs for other sectors, demand for outputs from other sectors, impacts on markets for substitutes. 2. Increases in factor productivity. 3. Economic and non-economic benefits—for the region and other regions—	1. Negative impacts on some sectors of inter-sectoral linkages from increased project outputs. 2. Decreases in factor productivity. 3. Indirect economic impacts of activities negatively affected by the project (e.g., negative multipliers of lost output from inundated area).

contd...

contd...

from migration into or reduced migration out of regions. 4. Dams construction may encourage watershed development actions aimed at reducing erosion/sediment concentration. These, in turn, have direct, indirect and induced impacts.	4. Economic and non-economic costs of migration into or reduced migration out of region (e.g., strain on local infrastructure and natural resource base). 5. Negative environmental impacts of economic activities indirectly affected via inter-sectoral linkages.
Induced (or Tertiary)	**Induced (or Tertiary)**
1. Positive impacts (regional/inter-regional/national) on output of increased income, expenditures for local/regional workers (either previously unemployed or employed at lower wages). 2. Impacts of increased remittances outside the region. 3. Increased property values due to reduced flood occurrences, water availability, recreational values, etc. 4. Income effect of reduced prices of goods/services provided by activities that are direct outcomes of the dam—benefits of non-marginal output changes. 5. Positive impact of increased government revenues (e.g., it reduces public debt burden). 6. Positive impacts of increased government expenditure.	1. Impacts of reduced income/expenditure of displaced people. 2. Impacts of reduced wages due to migration into the region, if it outweighs the impact due to increased demand for labour. 3. Increased public debt burden if project financed with public funds. 4. Negative impacts of increased government expenditure (e.g., crowding out). 5. Negative environmental impacts of increased incomes, changes in the structure of expenditure and changes in employment patterns.

Notes: (1) These are net of all costs of construction including the costs of noise, pollution, health impacts, injuries/deaths, temporary reduced value of area, etc., as well as all costs incurred in producing project outputs.

(2) In economic terms, the objective of flood alleviation or mitigation is to maximise the efficiency of use of the catchment as a whole: it is not about minimising flood losses nor solely concerned with the flood risk in one, perhaps a small, part of the flood plain of that catchment (Green *et al.*, 1993).

* GHG: Green House Gases.

To add to the complexity, while some of the impacts of the dam projects are 'direct', the others are 'indirect' with the definition of what constitutes 'direct' *versus* 'indirect' impacts varying greatly in the literature. In a number of studies, all the economic costs and benefits that were not specifically intended by project planners, as well as most non-economic impacts, are identified as 'indirect' impacts. However, this is a misnomer, as most of these unintended impacts should be seen as stemming 'directly' from the construction and subsequent use of the dam. This is the case, for instance, of social costs of resettlements, externalities from lost output and lost ecosystem functions due to inundation, or net benefits from initially unforeseen uses of stored water that emerge along the life of the project. In this sense, numerous discussions of 'indirect' impacts actually focus on extending the consideration of 'direct' impacts of dams to all the outputs directly affected by the project, and to more than one dimension (not just economic and financial, but also environmental, social, cultural, etc.). Table 1.1 attempts a classification of direct, indirect and induced benefits and costs that can stem from a projects' outcomes.

Given all this complexity, it is not surprising that no comprehensive assessment of the effects of dams has been done. As noted by the Operations Evaluation Department of the World Bank (1996), "irrigation projects have, in general, substantial benefits stemming from linkages between irrigation and other sectors of production. Thus, the increased income of irrigated farmers will result in increased demand of complementary inputs to production (fertilisers, tractors, fuel), increased scope for downstream processing of the irrigated crops, and increased spending on consumption goods produced locally and outside the region. Unfortunately, there are no estimates available on the indirect benefits of the projects under review."

Although the World Commission on Dams (WCD) considered various methodologies for estimating the indirect effects of dams in its Thematic Reviews, the main *WCD Report* (2000) does not report the estimates of multiplier effects or indirect impacts of the dams that were selected as case studies for the WCD. In the discussions on financial and economic performance of large dam projects, the economic performance has been evaluated in terms of economic internal rate of return (EIRR). However, in some of the case studies, indirect multiplier effects have been estimated. According to the WCD Case Study of Grand Coulee Dam and Columbia Basin Project (Ortolano *et al.*, 2000), there have been substantial multiplier effects from investments made in the basic sectors. According to the study, "these sectors generate between 1.5 to 1.7 dollars of total income within the local area for each dollar produced by the basic sectors." Total income from the basic sectors was of the order of US$617 million (in 1998 prices), while another US$309 million in non-agricultural income was generated. In the case of the WCD case study of Tarbela dam in Pakistan, it is mentioned that " it is beyond the scope of the Tarbela study to calculate the net economic impact of Tarbela on the economy", even though the study reports output multiplier of 2.39 for the agriculture (farming) sector (Anonymous, 2000).

In the chapter on "People and Large Dams: Social Performance", the *WCD Report* concludes that " [...] a simple accounting for the direct benefits provided by large dams— the provision of irrigation water, electricity, municipal and industrial water supply, and flood control— often fails to capture the full set of social benefits associated with these services. It also misses a set of ancillary benefits and *indirect economic (or multiplier) benefits of dam projects*" (*WCD Report*, 2000: 129, emphasis added).

The Study: Scope and Objectives

The aim of this study, supported by the World Bank, is to go further than previous attempts, including the *WCD Report*, and evaluate some of the above interactions, in particular the 'direct' and 'indirect' economic impacts of dams.

For the purpose of the present study, the 'direct' costs and benefits are all those that descend from the construction of the dam, as well as the benefits accruing to all beneficiaries of the water stored and other services provided by the structure and the impacts—positive or negative—to all populations, species and ecosystems affected either by the construction or by the changed regime of the river. 'Indirect' and 'induced' impacts are those that stem from the 'linkages' between the economic, environmental, social or other direct consequences of a dam project 'with the rest of the economy', as well as the non-economic impacts descending from such linkages. Among them are impacts due to changes in output and input use in sectors other than those affected directly by the dam, or changes in relative prices, employment and factor wages. The type and relevance of such indirect and induced impacts will depend on a number of issues, including the size of the project relative to that of the regional and national economy, its single- or multi-purpose nature, the level of inter relatedness between the regional and the national economy, and the source of financing, among others.

We illustrate further the distinction between direct and indirect impacts and their inter relationships. The major outputs of a multi purpose dam and reservoir project include—hydropower, irrigated agriculture, water supply, fishing, flood control, drought prevention and the value of recreational activities/tourism revenues. In addition, the irrigation canals of the project provide several non-irrigation benefits in terms of water for households, livestock, household enterprises, animals and birds, among others. Several additional direct outputs include navigation, tourism, reduction in floods etc. These direct impacts are measured in terms of additional agricultural output generated, hydropower generated, increase in tourism etc. In addition to these direct positive impacts these dams also produce a number of negative direct impacts which, inter alia, include social, economic, and cultural costs of resettlement; value of lost ecosystem/historical/cultural heritage sites; opportunity cost of lost output from inundated area and reduction in fish output upstream.

The major direct outputs from a dam in turn generate both:

(i) inter-industry linkage impacts, both backward and forward linkages, resulting in increase in the demand for outputs of other sectors, and

(ii) consumption-induced impacts arising out of increases in income and wages generated by the direct outputs of the dam.

These impacts are referred to as indirect impacts. Thus, the indirect and induced economic impacts are those impacts that stem from the linkages between the economic and other direct consequences of a dam project with the rest of the economy, as well as the non-economic impacts descending from such linkages. Among them are impacts due to changes in output and input use in sectors other than those affected directly by the dam, or changes in relative prices, employment and factor wages.

To illustrate, the hydropower produced from a multipurpose dam provides electricity for households in urban and rural areas and for increased output of industrial products (e.g., fertilisers, chemicals, machinery). Changes in the output of these industrial commodities require inputs from other sectors such as steel, energy, chemicals, among others. Similarly, water released from a multipurpose dam provides irrigation for increased output of agricultural commodities. Changes in the output of these commodities require inputs from other sectors such as energy, seeds, fertilisers, etc. Further, increased output of some agricultural commodities encourages setting up of food processing and other industrial units. Thus, both increased output of electricity and irrigation from a dam result in significant backward linkages (i.e., demand for higher input supplies) and forward linkages (i.e., providing inputs for further processing).

Increased outputs of industrial and agricultural commodities also generate additional wages and incomes for households. Higher incomes result in higher consumption of goods and services that, in turn, encourage production of various agricultural and industrial commodities. Further, changes in output generated by the project may affect prices of direct project outputs, inputs, substitutes, complements and factors of production. Changes in wages and prices have both income and substitution effects on expenditure and saving decisions of different owners of factors, which further impacts the demand for outputs both within the region and throughout the economy. Induced impacts reflect the feedbacks associated with these income and expenditure effects, and also include any impacts of changes in government revenues and expenditures that resulted from the project.

The present study thus aims to:

a. discuss the nature of some key indirect and induced economic, distributional and poverty reduction impacts of dam projects,
b. present the methodologies that are available for their assessment,
c. apply them to the *ex-post* evaluation of their magnitude in a few case studies, and
d. identify the most important lessons that can be drawn in relation to the relevance of multipliers or other measures of indirect and induced impacts for guiding investment decisions on dams.

The present study should be seen as one of the numerous steps that need to be taken to reach the goal of evaluating the full development impact of a dam project. The aim here is to highlight the relevance of one of the components of a full evaluation of dam projects that is often neglected, i.e., their indirect and induced 'economic' impacts. These have been estimated using different economy-wide, multi-sector modeling techniques for the case studies presented in Table 1.2. One of the major reasons for selecting these dams is that they have had significant impacts on the economies of the countries/regions where they are located, and at times on other regions as well. Their selection was also based on a consensus that these dams represented varying conditions in terms of affected sectors, structure of the regional economy, regional *versus* inter-regional *versus* national impact, and evolution of the policy environment to which they were exposed.

Table 1.2

Case Studies for Estimation of Indirect Economic Impacts of Dams

Sr No.	*Case Study Dam*	*Large/Small*	*Nature*	*Country*
1.	Bhakra Multipurpose Dam System	Large	Multipurpose	India
2.	Bunga Check Dams	Small (Check dams)	Single purpose	India
3.	The High Dam at Aswan	Large	Multipurpose	Egypt
4.	Sobradinho Dam and the Cascade of Reservoirs on the Sub-Médio São Francisco River	Large	Multipurpose	Brazil

The in-depth analysis of many other crucial impacts identified in Table 1.1—such as the evaluation of indirect environmental and social impacts—for selected dams, though no less important, have not been analysed in this study. However, in the context of the selected case studies, social and environmental impacts of each dam have been presented and discussed based on existing analyses and studies.

The Outline

The book is divided in to two parts. Chapter II reviews the available analytical tools and methodologies for estimating the direct, indirect and induced effects of dams. An overview of comparative results in respect of multiplier impacts of the four case studies are presented in Chapter III. The Chapter also gives an analysis of the income distribution impacts in respect of three of the four studied dams. Chapter IV essentially is written in the form of a guidance note for practitioners. Chapter V very briefly lists some further issues on which some more work needs to be done. Part II of the book describes in somewhat larger details the methodology and the results of all the four case studies. The details of Bhakra Dam system, Bunga Check Dams, Aswan High Dam and the Sobradinho Dam and the cascading reservoirs are presented in Chapters VI to IX. The last chapter in the form of "Afterword" describes the work reported in this book in a fine historical, political economic context.

References

Anonymous (2000). *Tarbela Dam and the Related Aspects of the Indus River Basin, Pakistan*. Case Study Prepared for the World Commission on Dams

Green, C.H., D.J. Parker and E.C. Penning-Rowsell (1993). "Desigining for failure", in P.A. Merriman and C.W.A. Browitt (eds.), *Natural Disasters: Protecting Vulnerable Communities*. London: Thomas Telford.

Ortolano, L., Cushing, K., and Contributing Authors (2000). *Grand Coulee Dam and the Columbia Basin Project, USA*. Case study report prepared as an input to the World Commission on Dams. Cape Town. *www.dams.org/report/*

WCD (2000). *Dams and Development: A New Framework for Decision-Making*. The Report of the World Commission on Dams. Issued on November 16, 2000. Available at *www.damsreport.org*

World Bank (1996). *The World Bank's Experience with Large Dams—A Preliminary Review of Impacts*. Operations Evaluation Department Report No. 15815. Washington D.C.: The World Bank.

2 Methodological Issues

Ramesh Bhatia and Monica Scatasta

Given the focus of this study on indirect economic effects and income distribution impacts of the dams, the key methodology issues are:

i. The estimation of multipliers or other measures that reflect production-related and consumption-induced indirect effects of dams, and the assessment of their regional, inter regional and/or national economic impacts.

ii. The assessment of income distribution and poverty reduction impacts.

iii. The use of estimates of indirect and induced impacts for *ex-ante* appraisal and *ex-post* evaluation of dam projects

Estimation of Multipliers

As discussed earlier, the direct outputs from a dam generate both inter-industry linkage impacts and consumption-induced impacts on the regional/national economy. The magnitude of indirect impacts of a dam on the regional output and value-added will however, depend on the strength of linkages amongst various sectors of the economy. Multiplier analysis offers one of the approaches for quantifying the magnitude of inter-industry linkages and consumption-induced effects in relation to purely direct impacts.

Multipliers are summary measures that reflect the total effects of a project in relation to its direct effects. A multiplier of 1.90, for instance, implies that for every 1 dollar of value-added generated directly by the project at maturity, another 90 cents are generated in the form of indirect or downstream effects. Thus, the multiplier is a ratio of the total effects (direct and indirect) of a dam project to its direct effects.

Estimation of multiplier requires careful analysis of the direct effects. This essentially involves quantification and valuation of major outputs of the dam and the assessment of the share of direct effects that is attributable to the dam project.

As shown in Table 2.1, for estimating a project multiplier value for a dam (e.g. the Bhakra dam), for the numerator, we need to estimate the regional value-added (say for the Punjab state) under 'With Project' situation as well as the regional value added under 'Without Project' situation. For the denominator, we need to estimate the value added from the sectors that are directly affected by the major outputs of the dam (namely agricultural output, hydro electricity, water supply etc.).

Table 2.1

Values of Variables Required for the Estimation of a Project Multiplier of the Bhakra Dam Project, India

Definition of Project Multiplier	Regional Value-Added with Project **minus** Regional Value Added without Project
	Value-Added of Agriculture and Electricity with Project **minus** Value-Added of Agriculture and Electricity without Project

Methodologies for the Estimation of the Indirect and Induced Impacts

A number of analytical tools have been suggested in the literature for estimation of multiplier effects (Bell *et al.*, 1982; Hazell and Ramasamy, 1991; Haggblade *et al.*, 1991; Hoffman *et al.*, 1996, and Aylward *et al.*, 2001). This section presents a summary review of these methodologies. The following are the key methods of multiplier analysis that have been considered:

1. Input-Output (I-O) and Semi-Input Output (S/I-O) Models.
2. Social Accounting Matrices (SAM) based models, including the IMPLAN Models.
3. Computable General Equilibrium (CGE) Models.

The analytical tools that can be used to analyse indirect and induced impacts of exogenous changes—be these demand shocks, policy changes, or the introduction of large projects with potential for such 'spillover' effects—fall into the broad category of multi-sector models. A continuum of such models exist, differing primarily with respect to (i) their assumptions regarding quantity *versus* price-responsiveness to exogenous shocks, (ii) the focus on income levels *versus* the inclusion of distributive considerations, (iii) their capacity to incorporate factor and import substitution possibilities, and (iv) their capacity to accommodate policy distortions, specific factor and output market structures, and other peculiarities.

Input-Output Models and Semi Input-Output Models

The core around which all economy-wide, multi-sector models are built is the input-output model pioneered by Leontief (Leontief *et al.*, 1953; Leontief, 1970). Input output analysis is a way to trace the flow of production among the sectors in the economy, through to final domestic or export demand. The essence of input-output analysis is that it captures the inter relatedness of production arising through the flow of intermediate goods among sectors. The fundamental input-output problem is that of a planner who wants to determine the appropriate adjustments in economic quantities throughout the economy in order to achieve a specific final output (see Hewings, 1985; Dervis *et al.*, 1982).

Input-output models are based on an accounting framework that records all inter-industry flows (at the level of disaggregation deemed relevant by the analyst), final demand by households, the government, an investment account and an export account, factor remuneration and total imports. Columns in an input-output table record payments from the column sector to the rows (other industries, factors, and imports), so that their total represents total gross inputs for a sector, while rows represent all receipts for a row sector from the columns (other industries and all final demand categories), so that their totals represent total gross output for the sector. Inter-industry linkages are based on a fixed coefficient square Leontief matrix, so that no substitution possibilities exist for producers. Similarly, factor remunerations represent fixed value-added proportions of total gross output.

Semi-input-output (S-I/O) models represent a variant of I/O models whereby a distinction is made between tradable and non-tradable goods. The former are assumed to have an exogenously set domestic level of output, so that any change in demand will be reflected in a change in exported quantities. In terms of domestic production, therefore, the whole brunt of demand shocks are borne by non-tradables. The implication of this distinction is that induced, consumption-based impacts will reverberate throughout the economy only via adjustments in non-tradables and their inter-industry linkages. This characterisation is important, in that it refines the representation of the specific regional structure of production, reducing the risk of overestimating induced impacts that are not felt by regional sectors.

Input-output and semi-input-output models evaluate indirect and induced economic impacts by computing multipliers that reflect the impact of a unitary change in one sector's output—due to an exogenous unitary change in final demand—onto the output of other sectors, income, and employment. Existence of a multiplier depends on drawing unused or underused resources into more productive economic activities (Haggblade *et al.*, 1991), so that the presence of such underutilised resources in the region of interest is crucial for the existence of multiplier impacts as estimated by this class of models. Leontief or output multipliers only reflect the degree to which industrial sectors are linked with each other and the strength of such linkages, but tell us nothing about the larger or smaller impact on a regional or national economy of increased demand for the output of any of those sectors. To do so, further manipulation is needed. Various multipliers have been estimated, but the classic examples are the so-called Type I and Type II income multipliers, where the first reflects direct and indirect economic impacts, while the latter also includes induced effects.

The main limitation of this set of models is that they assume linearity in production and cost-determined prices independent of demand. These are fundamentally planning-oriented models, based on the assumption that quantities on the supply side of the economy will respond to exogenous shocks on the demand side in the absence of any supply-side constraint. Supply functions are perfectly elastic (horizontal, instead of the more realistic upward-sloping shape reflecting the scarcity of one or more inputs), and both output and input prices are unaffected, regardless of the size of the exogenous shock. Haggblade *et al.* (1991) showed the dampening effect that an upward-sloping supply of non-tradables—linked either with the labour market or capacity expansion rigidities—can have on multipliers. These considerations are important when considering the agricultural or other sectors in developing countries, where such rigidities may be significant. Bell and Devarajan (1985), showed that the use of shadow prices based on a constant supply price of labour may be incorrect for large projects that affect wages. This was the case in their application of a semi-input-output model to the Muda River Scheme, which proposed the use of semi-input-output models as an alternative 'primal' route to social cost-benefit analysis, as opposed to the traditional 'dual' route of shadow prices.

Social Accounting Matrices (SAM) and SAM-based Multiplier Analysis

More recently, another set of linear models have been developed that extend the input-output table to include the distribution of income among 'institutions' (household categories, firms, the government), to better represent their expenditure, and to distinguish between production activities and produced commodities. The interest in income distribution and the flow of funds among institutions led to the development of a more general social accounting framework than the I/O-based 1968 System of National Accounts (SNA): the Social Accounting Matrix (SAM) (See Pyatt, 1994; 1991a; 1991b; 1985; Pyatt and Round, 1985).

Its structure can be translated into a set of linear equations with fixed coefficients, and used to compute fixed-price multipliers (Pyatt, 1988). Most of the limitations characterising I/O models—their demand driven nature, lack of price responsiveness, linear inter-sectoral interactions and factor use—also affect SAM-based multiplier analyses. However, they represent a step forward in that they close the loop between factor incomes, their distribution to different endogenous and exogenous economic agents, and their expenditure behaviour.

A Social Accounting Matrix (SAM) is a comprehensive, economy wide data framework that represents the circular flow of income and expenditure in the economy of a nation or region. Each cell represents a payment from a column account to a recipient row account. The SAM is a square matrix that explicitly traces the distribution of factor incomes to institutions. These institutions generally comprise numerous household groups—distinguished on the basis of income, location, etc.—firms and the government. Annex A.2.1 provides some more details on the nature and structure of SAMs, and on the characteristics of SAM-based models.

A crucial distinction between SAM-based multiplier analysis and I/O models is that SAMs are able to account for the way in which initial asset distribution and factor endowments—and therefore, the distribution of income among household categories and between them and the corporate sector—interact with the structure of production in determining final outcomes, particularly for welfare analysis. The key *raison d'être* for the SAM is that it embodies the fact that within the macro-economy there is a circular flow process and that what happens at one point on the circuit will have implications for experience at other junctures, so that one needs to be equally concerned with all the different aspects of technology and behaviour that together describe the circular income flow and the connections in the economy.

Each cell of the SAM can be seen as a 'block' representing sets of transactions. As rows and columns total must be equal, expressions linking row and column elements with their totals can be used to form a system of linear equations that embodies market, behavioural and system relationships. The first have to do with the accounts relative to goods and factor markets. Behavioural relationships regard the budget constraints of economic agents or institutions represented in the SAM. Finally, system relationships regard the capital and rest of the world accounts, where macroeconomic (internal and external) balances are represented. This system can be used to estimate multipliers.

Multiplier Estimates using SAM-based Models: The IMPLAN Models

IMPLAN Professional® is an economic impact assessment software system developed and marketed[1] by Minnesota IMPLAN Group (MIG). IMPLANPro, combined with databases maintained by the Minnesota

1. Over 500 clients across the country use the IMPLAN model, making the results acceptable in inter-agency analysis. The Natural Resources Inventory and Analysis (NRIAI) and Social Sciences Institutes (SSI) are supporting usage of IMPLAN throughout NRCS. They have provided IMPLAN training and models to each NRCS region and purchased all 50 state datasets. Any of over 2,000 county models can be obtained from MIG as needed. The IMPLAN software can combine any combination of counties and states into one study area.

IMPLAN Group, allows the user to develop local level input-output models that can estimate the economic impact of new firms moving into an area, professional sports teams, recreation and tourism, and many more activities. The IMPLAN models are available for all 50 states in the US and data for about 2000 counties are available that can be combined to estimate multipliers for any region in the country. Any research group can develop a complete set of social account matrices; develop case-specific multiplier tables; change any component of the system (production functions, trade flows, or database) and create impact analyses by entering final demand changes. Any user can obtain a report for a region that describes the structure of the economy in terms of the base year levels of employment, income, and output for all 528 IMPLAN sectors. The report also provides regional multipliers for output, income, other value added, and employment.

Computable General Equilibrium (CGE) Models

In a world where policy makers cannot control quantity variables directly—as it is implicitly assumed by linear I/O and SAM-based models—but rather affect them via the mediation of market incentives, one needs to understand how markets respond to different sets of policy interventions. CGE models are well suited to this purpose, providing a framework where endogenous prices and quantities interact to simulate the workings of decentralised markets and autonomous economic decision-makers. Following the pioneering efforts of Adelman and Robinson (1978) and Taylor *et al.* (1980), standard CGE models have been extensively used to study policy impacts on income distribution, growth and structural change in developing economies.[2]

A CGE model is a system of simultaneous nonlinear equations that provide a complete and consistent picture of the 'circular flow' in an economy, capturing all market-based interactions among economic agents.[3] Figure A.2.1 in Annex A.2.1 describes the circular flow of incomes for a simple open economy.

---

2. A concise description of a CGE model for developing countries is in Devarajan *et al.* (1990). Early reviews—Devarajan *et al.* (1986); Decaluwe and Martens (1988); Bandara (1991)—give an idea of the size of the literature.

3. See the Annex A.2.1 for details on the CGE models.

Four features distinguish CGE models from Leontief's input-output modeling tradition:

- Price endogeneity, as opposed to quantity adjustments, to reach an equilibrium.
- Price-responsive input and output substitutability—perfect or imperfect—through the use of nonlinear supply and demand equations (Robinson, 1989).
- The abandonment of the perfect dichotomy between traded and non-traded goods from traditional I/O models.
- Factor supply constraints, which generate output supply constraints.

Price endogeneity and general equilibrium interactions also set CGE models apart from their other major antecedent: economy-wide linear programming (LP) models.

Another relevant characteristic of CGE models is that neoclassical assumptions of perfect competition, perfectly functioning markets with flexible prices, and free products and factor mobility can be relaxed. Hence, they can be used to model special institutional features and market distortions—such as imperfectly competitive behaviour, quantity or price adjustment lags, and government interventions on prices or quantities. This enables researchers to assess the impacts of policies, projects or exogenous shocks in a second-best environment, and to compare their outcomes with the outcomes under first-best conditions. This feature is important from the point of view of our study in that the CGE models enable a comparison of the expected direct and indirect impacts of a dam project both in the context of the actual distortions present in the regional or national economy in which it operates, as well as in an undistorted context that is analogous to that assumed by the use of shadow prices for project evaluation.

Researchers at the International Food Policy Research Institute, have developed a standard CGE (Löfgren *et al.*, 2001), and the paper is available on the internet at *www.ifpri.org* The paper includes an equation-by-equation description of the model and of its required database. The paper discusses the implementation of the model in GAMS (the General Algebraic Modeling System) and is accompanied by a self-extracting zip file, which includes the GAMS files for the model, sample databases, simulations, solution reports, and a SAM aggregation programme. The Annex A.2.1 includes a detailed description of the model's structure, based largely on Löfgren *et al.* (2001).

Two general approaches exist for incorporating water resources in a CGE model, each best suited to study a different set of issues. The first approach considers water as a factor of production found in nature. Models that follow this approach have been used to study issues such as increasing water scarcity and the consequent changes in the scarcity value of water, uneven regional water distribution, and water allocation based on its scarcity value. Berck *et al.* (1991) present a CGE model of agricultural water use in the San Joaquin Valley, California. Robinson and Gelhar (1995) model arable land and water scarcity in Egypt to analyse the consequences of creating a water market and of introducing charges on agricultural water use. Mukherjee (1996) presents a single region CGE model for the Olifants river area in South Africa, to study inter-sectoral water allocation and its regional economic impacts.

A second approach is the one adopted by Decaluwé *et al.* (1997),[4] which considers water as a commodity produced by a regulated monopolist, sold on regulated markets at regulated prices, and used both as an intermediate input and as a final consumption good. The authors build a two-region model that specifies two different water production techniques whose parameters vary by region.[5] To reach an equilibrium, the model adjusts the output of the water production activities in the two regions, given water prices. This approach can be used to study the impacts of water pricing reforms to achieve cost recovery, but not to analyse scarcity issues and the value of water as a natural resource.

Choosing the Right Analytical Tool

From the point of view of the analysis of indirect and induced impacts of dam projects, the choice of analytical tool should not always favour the

4. A similar model, applied to a shared aquifer situation is Roe and Diao (1994).

5. Note that the spatial dimension is not meaningfully incorporated in this model as (i) there are no water exchanges between the two regions; and (ii) physical distances do not give rise to any transportation costs.

most sophisticated tool, but rather be driven by the assumptions regarding the mechanisms through which impacts are transmitted in the specific region of interest, and particularly regarding factor mobility.

When prices are assumed fixed, as in I/O models or SAM-based multiplier analysis, all adjustments occur through quantity changes. A change in demand for a sector's output results in changes in that sector's output supply and factor demand. In the absence of supply constraints, adjustments occur via impacts on labour or capital employment and inter-regional factor migration. The presence of idle labour or capacity somewhere in the system—either locally or in other regions, if the model is inter-regional—prior to the exogenous shock is thus crucial for the existence of quantity-driven multiplier impacts as estimated by these models.

On the other hand, a variable-price model, such as a standard CGE, implies the presence of supply constraints, so that for at least one factor the aggregate levels of factor employment are fixed. In this case, a change in sectoral demand results in relative price changes, thus determining substitution effects among inputs and among outputs, with factor relocation across sectors in the regional economy. CGE models, however, can be flexible enough to incorporate alternative specifications of factor mobility and wage differentiation, and have successfully been modified to solve as I/O models, so that comparisons can be made between the welfare impacts under alternative assumptions in this area (Hoffman *et al.*, 1996, Bautista *et al.*, 1999). The greater flexibility of CGE models therefore suggests their use whenever possible based on data and human resource availability.

Table 2.2 presents a summary view of various methodologies for estimation of multipliers and income distribution impacts along with their key assumptions.

Designing the Experiments and Interpreting the Results

Many of the outputs of the dam projects are not sold on well-functioning markets, so that their market price—if one even exists—does not reflect their actual value to society. A key issue in this context therefore, is the estimation of shadow prices for such outputs. A

Table 2.2

Multiplier and Income Distribution Analyses using Alternative Methodologies

Features / *Method*	*Quantity-induced Impacts*: *Type I Mult (Production, direct + indirect)*	*Type II Mult (Production & consumption, direct + indirect + induced)*	*Income Distrib.*	*Price induced Impacts (on output, income, and its distribution)*	*Assumptions on Production and Markets*	*Adjustment Mechanisms*
Closed Input-Output (I/O) Models	YES	YES	NO	NO	1. Fixed prices, demand-driven; 2. No supply-side constraints (assume unemployment) 3. Fixed production coefficients (constant returns to scale);	Adjustment through quantity changes
Semi-input-Output (I/O)	YES	YES	NO	NO	1-3. Same as above 4. Distinguish Tradables vs. Non-Tradables => only the latter produce induced impacts	Adjustments through quantity changes
Social Accounting Matrix (SAM) Models	YES	YES	YES	NO	1-3. Same as above 4. Semi-tradables possible both on side import and export 5. Closed income loop (explicit link between production structure & income distribution)	Adjustments through quantity changes
Computable General Equilibrium Models (CGE)	YES (if operated as SAM-based model)	YES (if operated as SAM-based model)	YES	YES	1. Relative price endogeneity; 2. Simulate both full employment and unemployment; 3. Can use non-linear production functions; 4. Accommodate imperfect substitution of domestic goods vs. imports, exports; 5. Closed income loop via factor price changes; 6. Simulates market and policy-induced distortions, can compare dam impacts under first- & second-best conditions	Adjustments in relative output and factor prices

number of possibilities arise. If the non-marketed output is used as an input by one or more beneficiary sectors—as is the case for water stored in the reservoir—one can use mathematical programming models of the basin or sub-basin to simulate reduced availability of water and their impacts, so as to estimate the shadow price of water associated with its availability constraint—for each sector and overall in the region or in separate river reaches or sub-basins. Often, however, the non-marketed output is multi-dimensional, as is the case for flood management, or some of the consequences of the presence of a dam will not be captured by a simple 'retiring of the waters'. In this case, appropriate proxies for the non-marketed impacts will have to be identified in variables whose value can be measured and can be changed exogenously to signify the introduction of the dam or simulate its absence.

Particular care has to be taken in designing the 'without dam' scenarios. First of all, these cannot be interpreted as full-fledged counterfactuals of the 'with dam' case. A CGE model of a region where a dam is located will be based on a SAM that replicates the structure of the economy, factor distribution, income flows, etc. that have evolved in part because of the introduction of the dam project in the past. Path dependance is inescapable. Hence, the simple modification of a few parameters or exogenous modification of initial variable levels will not fully capture the alternative configuration that the region's economy would have had, had an alternative development path been chosen. The capacity of designing meaningful 'without dam' scenarios is therefore crucial in the quality of the insights that such a study is able to provide.

Defining the Region's Boundary

In studies on the multiplier effects of dams where a regional perspective is considered important, there arises the difficult problem of defining the region that surrounds the project or is affected by the project. In some cases such as the Aswan, Mahaweli and Bhakra dams, the impacts of the project may go beyond the surrounding region and may affect incomes and consumption levels in far away regions. For example, in the case of the Bhakra dam located in the Punjab state in India, irrigation, drinking water and power benefits are available to four other states—

Haryana, Rajasthan, Delhi and Himachal Pradesh. High demand for agricultural labour in the Bhakra command areas attract migrant labour and the incomes transferred by these labourers to their villages in Bihar and eastern Uttar Pradesh result in investments in water and sanitation facilities hundreds of miles away. Further, the Bhakra dam contributes significantly to meeting the foodgrain requirements of urban poor in major cities in India. In the year 2000, these two states provided 17 million tonnes of wheat and 7 million tonnes of rice to the central pool to be distributed through fair price shops to the urban poor. Thus, the impact of the surplus foodgrains from the Bhakra Dam are often felt by people in areas far away from the location of the dam.

When analysing a dam project, therefore, the model structure and related database should present the appropriate regional characterisation so as to incorporate as much of indirect effects as possible. These studies will have to be supplemented by other evaluations that might have been carried out in other studies (e.g. the impact of the Bhakra dam on food prices for the urban poor and the impact of remittances in villages in Uttar Pradesh and Bihar).

Analysis of Income Distribution and Poverty Reduction Impacts of Dams

As shown in Table 2.2, SAM-based models and CGE models explicitly take income distribution considerations into account by disaggregating households into various income categories.

In order to carry out a meaningful welfare and poverty impact analysis of the consequence of a dam project, it is important to design a CGE model so that the groups that stand to lose or gain from the project are clearly identified, and that the direct and indirect channels through which such gains and losses take place are defined. de Janvry and Sadoulet (2001), emphasise the need to characterise clearly the specific structure of poverty in the region/country of interest, and characterise the sources of income and structure of consumption for the rural and urban poor. To evaluate income multiplier effects in agriculture, for example, it is crucial to understand the incidence of off-farm income and the access to labour markets of workers otherwise utilised in family farm work, as well as the access of family workers to labour markets.[6]

Factors Affecting the Level of Multiplier Values

A number of case studies will need to be undertaken to assess the relative importance of various factors that influence the level of multiplier values. Amongst the factors that may affect the level of multiplier values may include: (i) the size and type of the dam—single-purpose *versus* multi-purpose; (ii) single-crop *versus* multi-crop agriculture; (iii) level of increase in outputs and incomes and consumption propensities; and (iv) policy and institutional regimes influencing incentives for input-use and crop yields

It is expected that the level of multiplier values will be higher for a large multi-purpose dam compared with those for a single-purpose dam. Outputs of irrigated agriculture, hydropower, recreation will create substantial backward and forward linkages and consumption-induced indirect effects. It is also expected that the level of multiplier values will be higher for a large dam providing water for multiple-crops compared with those for a single-crop system (e.g. Muda project with paddy monoculture). For example, if outputs include oilseeds, sugarcane and cotton, there will be investments in agro-processing units, sugar mills and cotton ginning and processing mills and/or textile units.

---

6. The authors also importantly note that while the choice of numeraire does not affect any real value obtained from the simulation, it can affect the identification of impacts as direct vs. indirect, and may therefore affect the policy implications that can be derived from the model. In particular, they state that "Technological change in agriculture is expected to induce a decline in agricultural prices relative to non-agricultural prices, in particular because of the increased demand for other domestically produced goods due to income effects. If, at one extreme, the numeraire is an index of non-agricultural prices, then the relative price change is read as a decline in the agricultural price, leading to important positive indirect effects on real incomes and probably very low or even negative direct profit effects in the agricultural sector. If, on the other hand, the numeraire is close to the agricultural prices, relative price changes are read as an increase in the non-agricultural prices, leading to high direct effects and negative indirect effects. To seek a more neutral choice, we use as numeraire the nominal exchange rate, so that all results can be read in dollar terms, and the decomposition of effects [...] reflects changing prices relative to the dollar price of the commodities."[p.5] Hence, these impacts of the numeraire choice should be carefully assessed when designing a CGE model for the case studies.

Policy and institutional regimes influencing incentives for input-use and crop yields will include land tenure system, subsidies for inputs (water, fertilisers, HYV seeds) and support prices for crop outputs. Subsidies for inputs may encourage higher levels of the use of these inputs resulting in higher crop yields. Support prices provide incentives for higher input-use and yields and higher levels of marketed surplus.

The level of multipliers will also differ during different phases of the implementation of a dam project. During the construction phase, the project will require construction materials generating backward linkages with various supplier industries. During the construction phase, large numbers of workers will be engaged, some of whom would remit funds to other regions of the country. The multiplier levels will be relatively high during the phase when full benefits of the project are available since both the inter-industry linkages and consumption-induced effects will be quite significant. A number of case studies will have to be undertaken to assess the relative importance of various factors that influence the level of multiplier values.

The Use of Multipliers for Project Appraisal and *Ex-post* Evaluation

A question that needs to be resolved is which of the economic impacts should be included in the project appraisal and which in *ex-post* evaluation, and for what purposes. *Ex-post* dam project appraisal tends to focus solely on the value-added produced by the sectors directly affected by the project, with no consideration for the 'multiplier effects' their outcomes produce on the regional, national, and sometimes global economy.

The broad principles and applications of the use of shadow prices in project evaluation are available in a number of studies. As suggested by Gittinger (1982), for final goods and services, the shadow price is the value in use. For intermediate goods and services, the shadow price is the opportunity cost, the benefit foregone by using a scarce resource for one purpose instead of its next best alternative use. Opportunity cost, for intermediate goods and services is equal to the marginal value product, the value of additional output generated by an additional unit of variable input. For imports and exports, the c.i.f. or f.o.b. prices are adjusted by the shadow price of foreign exchange that reflects the foreign exchange premium. The shadow wage rate may be lower than the regulated wage rate if the labour comes from a region with relatively surplus labour supply.

The use of shadow prices, instead of market prices for major outputs and inputs has significant implications for the estimation and use of multipliers for project evaluation. It has been suggested by Gittinger that the multiplier analysis is generally not used in project analysis because of the difficulty of avoiding double counting of benefits. Shadow prices that include carefully traced indirect changes in value-added include the multiplier effects while minimising the danger of double counting. According to Gittinger, the stemming-from benefit is a form of indirect benefit that accrues to purchasers of project outputs that are themselves intermediate products. The direct and indirect value-added generated in forward-linked industries would exist, for example, if cotton that is produced by an agricultural project would be used in existing cotton gins having unused production capacity might allow those gins to create additional value-added that would not exist without the project. The stemming-from benefits would not normally exist where project outputs are traded, since the same stemming-from benefit could be had by increasing imports of the intermediate product or by diverting exports to local use.

Two issues need to be resolved. The first is whether the approaches currently used by practitioners of project analysis use shadow prices such that the direct and indirect effects of the project are incorporated in the analysis. The project evaluation literature provides rules for deriving shadow prices (that is social scarcities) of all goods and factors, which are applied solely to the direct inputs and outputs of a project. Do these shadow prices capture all downstream effects, too? For example, in the case of the Muda project, "what becomes of the 83 cents per dollar that are generated downstream for every dollar of value-added generated directly by the project?"[7]

7. Bell *et al.* (1982: 11).

This issue has been discussed in a number of papers. It has been pointed out that "the road to allowing for indirect effects in an L-M (Little-Mirrlees) framework is not without pitfalls. Moreover, since Little and Mirrlees (1974) do not specify in detail how these effects are to be incorporated into the estimation of shadow prices, it is not surprising that different interpretations arise" (Bell and Devarajan, 1980). And yet, the separate consideration of multiplier impacts has traditionally been excluded from social cost-benefit analysis (CBA) based on the argument that, for the purpose of estimating the welfare changes produced by a new project, the use of shadow prices to value project outputs is sufficient given certain standard assumptions (see Little and Mirrlees, 1974, Squire and van der Tak, 1975, Scott, 1976; Ward and Deren, 1991).

Both in principle and when applied to the stream of direct inputs and outputs associated with a project, the shadow prices used in a full cost-benefit analysis account fully for all downstream effects because they contain, among other things, all the information obtainable from a semi-input-output analysis. In practice, however, shadow prices are not usually calculated on the basis of such complete information, but are derived using the shortcut methods proposed in the standard literature. Further, it is stated that "capturing multiplier effects in cost benefit analysis requires, in principle, that the estimates of the relevant parameters pertain to the region in question. Since developing countries are usually regionally heterogeneous and input-output tables are not available at the regional level, the cost-benefit analysis usually may not capture downstream effects."

In addition, two of the standard assumptions that are deemed sufficient for the comprehensiveness of shadow prices are of particular interest, as these are likely to be violated in the case of dam projects in developing countries. The first key assumption justifying the exclusion of multiplier effects from consideration when assessing societal cost and benefits of projects is that shadow prices suffice under full employment conditions. When this condition does not hold, as may well be the case in a number of agricultural and industrial sectors in developing countries where dams are introduced, multiplier (secondary) impacts on output, income and employment can indeed be large and should not be ignored. In fact the World Bank Operational Manual Statement (May

1980) on "Economic Analysis of Projects" does state that where specific excess capacities may exist, then the effects of different patterns of second-round expenditures out of incomes generated by the projects should be allowed for in the estimation of efficiency (economic) prices. The Manual, however, does not suggest any methods for this purpose, nor does the *Handbook on Economic Analysis of Investment Operations*.

Even if one assumes full employment, a second assumption behind the use of shadow prices for social CBA is for a project not to alter currently prevailing relative prices. While contingencies for price uncertainties are included in the forecasting of conditions over the life of the project, these are not based on any conceptual framework capable of tracing the potential impacts of a large project on relative prices in the region or country. When a project is expected to have large enough impacts so as to affect input or output prices, such a framework is needed in order to capture the full range of consequences of the project. In addition, as market distortions are progressively eliminated in a growing number of developing countries, fluctuations in relative market prices have become a more preeminent preoccupation for the determination of project viability (Harberger, 1996b; 1996c). In this context, the evaluation of large projects with marketed outputs could greatly profit from the use of a conceptual framework that is able to trace the response of relative prices to exogenous shocks and the adjustments of the regional or national economy to relative price changes. Computable general equilibrium models are a tool of choice, as they simulate how decentralised decision-makers in production and consumption react to policy changes or exogenous shocks so that a new equilibrium position is reached through the adjustment of relative prices in the context of a SAM-based circular flow of income in the economy.

Thus, when standard assumptions behind traditional CBA methods are not valid, these may not fully capture project net benefits.[8] In addition, even where these assumptions are valid, regional and

8. In addition, social CBA methods suggested in the literature are seldom followed to the letter, as discussed in Little and Mirrlees (1991). Additional references on CBA methods and applications are: UNIDO (1972), Sinha and Bhatia (1982) and Young (1996; 2001).

macroeconomic impacts may have potentially large consequences in terms of income distribution, spatial configuration of regional development, and non-economic impacts on regions other than the project area. These are often of great interest to project developers or to the government of the region or country where the project is located. While not needed as part of a project's social CBA, these impacts need to be adequately evaluated and presented to decision-makers, so that they can make an informed decision based on the broader set of consequences produced by a project and its alternatives.

The "WCD Thematic Review on Financial, Economic and Distributional Analysis" states the distinction between CBA and economic impact analysis, as follows: "Whereas cost-benefit analysis is concerned exclusively with comparisons of benefits and costs to society created by a dam, economic impact analysis examines the distribution of the full range of economic impacts and outcomes that may occur as the result of a project, policy or other intervention." (Aylward *et al.*, 2001). The document clearly discusses the nature of indirect (secondary and tertiary) economic impact, the methodologies available for their assessment, and the ways in which their consideration should be integrated in *ex-post* cost-benefit economic project analysis.

Its conclusions are that:

"[...]regional and macroeconomic impacts should be undertaken in the following instances:

- when standard assumptions of full employment of resources are not fulfilled;
- when there is risk that secondary market effects may have important negative consequences for the macro economy or vulnerable groups and, hence, have implications for macroeconomic policy and poverty alleviation;
- when projects have explicit macroeconomic goals, such as the earning of foreign exchange from hydropower exports; or
- when projects have explicit distributional goals, such as the stimulation of regional economic development."[9]

---

9. Aylward (2001).

For this purpose the WCD report suggests the use of one or more tools, including input-output, social accounting matrix-based, and computable general equilibrium models. These are proven, valuable techniques for the assessment of indirect and induced impacts of projects, policies or exogenous shocks that are not immediately apparent—and at times counter intuitive.

One of the main concerns expressed in the WCD document regards the limitations of such models as predictive tools. The WCD report is especially concerned with predictive capacities because its focus is on *ex-ante* analysis for project selection. *Ex-post* analysis is not explicitly considered. The use of the above-mentioned analytical tools as *ex-post* evaluation tools, however, might greatly diminish these concerns. It is considered that valuable lessons can be learned by assessing the full range of development impacts of dam projects in the context of *ex-post* evaluations, and that *ex-post* evaluation can be a valuable tool for the identification and design of alternatives for new projects, as well as for their *ex-post* assessment.

A Brief Review of Some of the Past Studies on Multipliers

Having discussed the methodological issues in estimating the direct and indirect effects of dams, we present a review of some of the available studies in which multipliers have been explicitly estimated. As mentioned earlier, there are only a few studies of this sort for developing countries. One such study is a comprehensive analysis of the Muda dam project in Malaysia. The direct and indirect or downstream effects of the Muda dam project have been analysed in a very comprehensive study (Bell *et al.*, 1982). Other studies discussed below include[10]: Grand Coulee Dam in the Columbia basin, USA, and water re-allocation in the Klamath Basin in USA.

10. There are a number of other studies where multipliers have been estimated but these are not for a specific dam project as such. These include, *inter alia*, Hazell and Ramaswamy (1991), Bouhia (2001), Bhalla *et al.* (1990).

Multiplier Analysis for the Muda Dam in Malaysia

The Muda project in the northern peninsula of Malaysia was designed to bring around 100,000 hectares of paddy land under double-cropping to achieve food self-sufficiency in the country. The project that includes Muda and Pedu dams and their reservoirs (with a storage capacity of about 800,000 acre-feet) and two main canals and distribution systems, was fully operational by 1974. By 1974 total outlays on the capital account had reached Malaysian (M) $270 million, of which about M$100 million was financed by a loan from the World Bank to cover the project's direct foreign exchange cost.

In the case of the Muda dam[11], a multiplier of 1.83 was estimated, showing that for every one dollar of value-added generated directly by the project at maturity, another 83 cents were generated in the form of indirect or downstream effects. It was estimated that, in the aggregate, the project induced an increase of M$110 million in regional value-added. Of this, M$60 was due to the increase in paddy output and M$50 million was due to indirect or downstream effects. Indirect or downstream effects are further disaggregated into two components: (i) production linkages between agriculture and the rest of the economy, and (ii) consumers expenditure linkages. For the Muda project in Malaysia, it was further estimated that for every dollar of value added generated directly by the project, only 33 cents of the additional indirectly generated value-added descended from operation of production linkages, compared with 83 cents when both production and expenditure linkages are allowed for.[12] This shows that induced consumption (income) impacts need to be considered as these represented 50 out of the 83 cents of indirect impacts. Table 2.3

11. According to the analysis of sources of change using SAM models, the Muda project accounted for 56 per cent of the total increase in regional gross output (of M$199 million) between 1967 and 1974. The project accounted for 48 per cent of the total increase in regional net value added.

12. See Bell *et al.* (1982: 179). The authors conclude that "In the case of this project, then, the omission of consumers' expenditure linkages, which interact with the system's interindustry linkages, would lead to a gross underestimate of the project's downstream effects. Further, the study also estimates the effect of the project on distribution of incomes in the region and outside the region. It was estimated that the project accounted for about 80 per cent of the increase in the real incomes of paddy farmers and 129 per cent increase in the real incomes of landless paddy farm workers."

presents the two basic types of multipliers that can be used to assess a dam's indirect and induced impacts, with examples of their estimated values for the Muda project.

Table 2.3

Estimated Values for Multiplier Effects of the Muda Project, Malaysia

	Type I Multiplier	*Type II Multiplier*
Definition	Direct Effects + Production Linked Indirect Effects / (Direct Effects)	Direct Effects + Production Linked Indirect + Effects Consumption Expenditure-Induced Indirect Effects / (Direct Effects)
Value (Million $ Malaysian)	$\frac{60 + 19.6}{60} = 1.33$	$\left[\frac{60 + 19.6 + 30.4}{60} = 1.83\right]$

Source: Bell *et al.* (1982)

Multiplier Analysis for Grand Coulee Dam, USA

The Grand Coulee Dam (GCD) and the Columbia Basin Project (CBP) is one of the case studies undertaken by the World Commission on Dams (WCD). The first stage, which consisted of the dam and two powerhouses, started in 1933 and was completed in 1951. The second stage consisting of the "Third Power plant" was completed in 1975. The GCD has a total generating capacity of 6809 MW, is the largest producer of electricity in the US and during the 1990s, gross generation in some years was greater than 25 billion kWh.

Construction of CBP began in 1945 and the last addition of lands in the CBP occurred in 1985. Only about half of the 416,000 ha proposed for the CBP are receiving irrigation water because the "second half" of project lands were never developed. Gross value of output per acre (in constant dollars) doubled from 1962 to 1992. If the predicted crops had been grown in the expected proportions, the 215,000 ha irrigated in 1992 would have produced approximately US$338 million in 1998. The actual cropping pattern (high-value fruits and vegetables) resulted in a gross value of production of US$637 million (in 1998 prices), nearly twice as much as predicted.

A comprehensive analysis of beneficiaries prior to the decision to build the Grand Coulee Dam was made in 1932. This report identified direct benefits in the form of profits to farmers and businesses, indirect benefits in the form of increased values for farmland and urban land, and public benefits in the form of increased supply of produce for consumption and processing as well as reduced food prices. The 1932 report also recognised that investments generated by backward and forward linkages from direct investments, i.e. the increased economic activity for agribusiness, banking, wholesaling, retailing, and transportation, etc. should also be counted as project benefits and should be asked to contribute to cost recovery. In this regard, the Butler Report argued that "benefits to local business interests (were) precisely the same character as benefits to the farmer" (USACE, 1933: 1042).

An *ex-post* analysis conducted by the WCD revealed that the multiplier effects from irrigation predicted qualitatively in 1932 indeed have occurred (Ortolano *et al.*, 2000). According to a 1996 study (Olsen, 1996), it is found that the so-called "basic sectors" of agriculture, agricultural services, and food processing, account for 30 per cent to 50 per cent of all income in the counties in which the CBP is located. There have been substantial multiplier effects from investments made in the basic sectors. According to the study, "these sectors generate between 1.5 to 1.7 dollars of total income within the local area for each dollar produced by the basic sectors." Total income from the basic sectors was on the order of US$617 million (in 1998 prices), while another US$309 million in non-agricultural income was generated. The multiplier value will be somewhat higher if agricultural services and food processing are included as indirect/downstream effects.

At the state level, the Columbia Basin Project provided similar indirect benefits. These included an employment multiplier of between 1.4 and 2.5 jobs generated per direct agricultural industry job, and about 5.4 jobs per food processing job.

It may be mentioned that the direct and indirect benefits from power generation from the Grand Coulee Dam (GCD) have not been quantified. The GCD contributed about 30 per cent of the total power output in the region and, according to the Bonneville Power Administration (BPA), on a cumulative basis, by the early 1990's, the

GCD had generated revenues equivalent to over US$2.9 billion (in nominal dollars).

The case study also cites the estimates of capital gains termed as "assessable assets" that have resulted from the implementation of the CBP. One such benefit of the CBP is its effect on land value, which has substantial local benefits. For the total area currently receiving CBP water (276,500 ha), this represents an aggregated increased value of US$574 million (in 1998 prices).

Multiplier Analysis for Water Reallocation in Klamath Project, USA

In a recent study on the impact of water re-allocation, it was estimated that reduced water deliveries to agricultural operations in the Klamath Project—the so-called Klamath Irrigation Project Operations Plan (KPOP)—which was prompted by drought and endangered fish protection, may have significant economic impact on agriculture and the regional economy of the Upper Klamath Basin (Oregon State University, 2001). Based on a three-county IMPLAN model, it was estimated that the KPOP can be expected to reduce agricultural output by 30 per cent in the short run. The direct impact on output from the Klamath Irrigation Project (KIP) on agriculture and associated processing was estimated at US$95 million, and farmers are expected to lose about $46 million in income.

The impact will be felt in other sectors of the economy because of the economic interrelationships between sectors. The KPOP is expected to generate US$22 million in indirect output reductions for industries that supply the directly affected sectors, and a US$17 million reduction in spending by local households employed in the directly and indirectly affected sectors. Overall, the KPOP is thus expected to reduce output by US$134 million and incomes by US$68 million, costing the Klamath Basin more than 2,000 jobs, about 3.5 per cent of its total. The expected multiplier effects (short run, for one year) of reduced water supplies for irrigation under the KPOP are shown in Table 2.4.

Table 2.4

Estimated Values for Multiplier Effects of Reduced Water Supplies for Irrigation in the Klamath Basin

	Type I Multiplier		Type II Multiplier		
Definition	Direct Effects +	Production Linked Indirect Effects	Direct Effects +	Production Linked + Indir. Effects	Consumption Induced Effects
	(Direct Effects)		(Direct Effects)		
Income or Value-Added Multiplier (Million US$)	$\frac{46+12}{46} = 1.26$		$\frac{46+12+10}{46} = 1.5$		
Output Multiplier (Million US$)	$\frac{95+22}{95} = 1.2$		$\frac{95+22+17}{96} = 1.4$		
Employment Multiplier (No. of Jobs)	$\frac{1356+430}{1356} = 1.3$		$\frac{1356+430+270}{1356} = 1.5$		

Notes: The figures are for reductions in income, output and employment and show direct, indirect and induced (by consumption reduction) effects of reduced water deliveries to agricultural operations in the dry year, 2001.

Source: Oregon State University (2001)

References

Adelman, I. and S. Robinson (1978). *Income Distribution Policy in Developing Countries*. California: Stanford University Press.

Armington, Paul S. (1969). "A Theory of Demand for Products Distinguished by Place of Production" *IMF Staff Papers* Vol. 16.

Arndt, Channing, A. Cruz, H. Tarp-Jensen, S. Robinson and F. Tarp (1998). "Social Accounting Matrices for Mozambique, 1994, 1995", *Trade and Macroeconomic Division Discussion Paper* No. 28. Washington D.C.: IFPRI

Aylward, B., J. Berkhoff, C. Green, P. Gutman, A. Lagman, M. Manion, A. Markandya, B. McKenney, K. Naudascher-Jankowski, B. Oud, A. Penman, S. Porter, C. Rajapakse, D. Southgate and R. Unsworth (2001). *Financial, Economic and Distributional Analysis*, Thematic Review III.1. Prepared as an input to the World Commission on Dams. Cape Town. *www.dams.org*

Bandara, J.S. (1991). "Computable General Equilibrium Models for Development Policy Analysis in LCDs", *Journal of Economic Surveys* 5(1): 3-69.

Bautista, R.M., S. Robinson, and M. El-Said (1999). "Alternative Industrial Development Paths for Indonesia: SAM and CGE Analyses", *TMD Discussion Paper* No. 42. Washington D.C.: IFPRI

Bell, C. and S. Devarajan (1985). "Social Cost-Benefit Analysis in a Semi-Input-Output Framework: An Application to the Muda Irrigation Project", in G. Pyatt and J.I. Round (eds.), *Social Accounting Matrices: A Basis for Planning*. Washington D.C.: The World Bank..

Bell, C , P. Hazell, and R. Slade (1982). *The Project Evaluation in Regional Perspective—A Study of an Irrigation Project in Northern Malaysia.*

Bell, C and S. Devarajan (1980). "Semi-Input-Output and Shadow Prices: A Critical Note", *Oxford Bulletin of Economics and Statistics.* August.

Berck, P., S. Robinson, and G. Goldman (1991). "The Use of Computable General Equilibrium Models to Assess Water Policies", in Ariel Dinar and Zilberman (eds.), *The Economics and Management of Water and Drainage in Agriculture*. Boston: Kluwer Academic Publisher.

Bhalla, G.S., G.K.Chadha, S.P. Kashyap and R.K.Sharma (1990). *Agricultural Growth and Structural Changes in the Punjab Economy: An Input-Output Analysis.* Washington D.C.: IFPRI

Bouhia, H. (2001). *Water in the Macro Economy*. London: Ashgate Publishing

Buckely, Patrick H. (1992). "A Transported Oriented Inter-regional CGE Model of the United States", *The Annals of Regional Science*, 26(4).

Brooke, A., D. Kendrick and A. Meeraus (1988). *GAMS: A User's Guide*. San Francisco, CA: The Scientific Press.

Decaluwé, B. and A. Martens (1988). "CGE Modeling and Developing Countries: A Concise Empirical Survey of 73 Applications to 26 Countries", *Journal of Policy Modeling* 10: 529-568.

Decaluwé, B., A. Patry, and L. Savard (1997). *"Quand L'eau n'est plus un don du ciel: Un MECG appliqué au Maroc"*, *Cahier de Recherche* No. 97-18. Canada: Département d'Economies, Université Laval.

de Janvry A. and E. Sadoulet (2001). *World Poverty and the Role of Agricultural Technology: Direct and Indirect Effects.* Draft.

Dervis, K., J. Demelo, and S. Robinson (1982). *General Equilibrium Models For Development Policy*. Cambridge, UK: Cambridge University Press.

Devarajan, S., J.D. Lewis and S. Robinson (1994). *Getting the Model Right: The General Equilibrium Approach to Adjustment Policy.* Unpublished manuscript.

———. (1990). "Policy Lessons from Trade-Focused, Two-Sector Models", *Journal of Policy Modeling* 12(4): 625-657.

———. (1986). "A Bibliography Of Computable General Equilibrium (CGE) Models Applied To Developing Countries", *Working Paper* No. 400. Berkeley: Division of Agriculture and Natural Resources, University Of California.

Dixon, P.B., B.R. Parmenter, A.A. Powell, and P.J. Wilcoxen (1992). *Notes and Problems in Applied General Equilibrium Economics*. New York: North-Holland.

Elberno, Chris T.M. (1995). "Linking CGE Models: Modeling the Transport Sector and Spatially Homogenous Goods", in J.C.J.M. van den Bergh *et al., Recent Advances in Spatial Equilibrium Modeling: Methodology and Applications*. Springer Verlag

Elbers, C. (1995). "Linking CGE Models: Modelling the Transport Sector and Spatially Homogenous Methodology and Applications", in J. vander Bergh, C.J.M. Jeroen, P. Nijkamp and P. Rietvold (eds.), *Recent Advances in Spatial Equilibrium Modelling: Methodology and Applications*. New York: Springer Verlag.

Ginsburgh, V., and M. Keyzer (1997). *The Structure of Applied General Equilibrium Models*. Cambridge, Mass., USA: MIT Press.

Ginsburgh V. and S. Robinson (1984). "Equilibrium and Prices in Multisector Models", in M. Syrquin, L. Taylor and L. Westphal (eds.), *Economic Structure and Performance*. New York: Academic Press.

Ginsburgh, V. and J. Waelbroeck (1981). *Activity Analysis and General Equilibrium Modeling*. Amsterdam: North Holland

Gittinger, P. (1982). *Economic Analysis of Agricultural Projects*. 2nd Edition. EDI Series in Economic Development, Baltimore, MD: The Johns Hopkins University Press.

Haggblade, S., J. Hammer, and P. Hazell (1991). "Modeling Agricultural Growth Multipliers", *American Journal of Agricultural Economics* 73(2): 361-74.

Harberger, A. (1996a). "Reflections on Social Project Evaluation", in D.G. Davies (ed.) *The Economic Evaluation of Projects: Papers from a Curriculum Development Workshop*. EDI Learning Resources Series. Economic Development Institute and Washington D.C.: The World Bank

————. (1996b)." Project Evaluation for the Next Decade", in D.G. Davies (ed.) Ibid.

————. (1996c). "Notes on Some Issues in Social Project Evaluation", in D.G. Davies (ed.) Ibid.

Harris, B., P. Hazell, and R. Slade (1987), "Regional Growth Linkages from Agriculture", *Journal of Development Studies* 23(2): 274-94.

Hazell, Peter B.R. and C. Ramasamy (1991). *The Green Revolution Reconsidered: The Impact of High-Yielding Rice Varieties in South India*. Baltimore, MD: The Johns Hopkins University Press.

Hewings, G.J.D. (1985). *Regional Input-Output Analysis*. California: SAGE Publications, Beverly Hills

Hoffman, S., S. Robinson, and S. Subramanian (1996). "The Role of Defense Cuts in the California Recession: Computable General Equilibrium Models and Interstate Factor Mobility", *Journal of Regional Science* 36(4): 571-95.

Leontief, W. (1970). "The Dynamic Inverse", in Carter and Brody (eds.), *Contribution of Input-Output Analysis*. Amsterdam: North Holland

Leontief, W. *et al.* (1953). *Studies in the Structure of the American Economy*. New York: Oxford University Press.

Little, I.M.D. and J.A. Mirrlees (1991). "Project Appraisal and Planning Twenty Years On", in *Proceedings of the World Bank Annual Conference on Development Economics 1990*. Washington D.C.: The International Bank for Reconstruction and Development/The World Bank.

———. (1974). *Project Appraisal and Planning for Developing Countries*. London: Heinemann

Löfgren, Hans (2000a). "Exercises in General Equilibrium Modeling Using GAMS". *Microcomputers in Policy Research* 4a. Washington D.C.: IFPRI.

———. (2000b). "Key to Exercises in General Equilibrium Modeling Using GAMS", *Microcomputers in Policy Research* 4b. Washington D.C.: IFPRI,

———. (1996). "The Cost of Managing With Less: Cutting Water Subsidies and Supplies in Egypt's Agriculture", *Trade and Macroeconomic Division Discussion Paper* No. 7. Washington D.C.: IFPRI

Löfgren, H., Harris, R.L., and S. Robinson (2001). "A Standard Computable General Equilibrium (CGE) Model in GAMS", *Trade and Macroeconomics Division Discussion Paper* No. 75. Washington D.C.: IFPRI,

Löfgren, H. and S. Robinson (1999). "Spatial Networks in Multi Region CGE Models", *Trade and Macroeconomic Division Discussion Paper* No. 35, Washington D.C.: IFPRI.

———. (1997). "The Mixed Complementarity Approach to Specifying Agricultural Supply in CGE Models", *Trade And Macroeconomic Division Discussion Paper* No. 20. Washington D.C.: IFPRI.

Mukherjee, N. (1996). "Water and Land in South Africa: Economy-Wide Impacts of Reform", *Trade and Macroeconomic Division Discussion Paper* No. 12. Washington D.C.: IFPRI

Olsen., D. (1996). *The Columbia Basin Project: Project Operations and Economic Benefits —A Regional Overview*. Report prepared for the Pacific Northwest Project.

Olson, Carl. E and Odell L. Walker (1996). "Agricultural Adjustments and Area Economic Interrelationships", *Journal of Farm Economics* 48(5). December.

Oregon State University (2001). *Economic Impacts of 2001 Klamath Project Water Allocation*. Department of Agricultural and Resource Economics.

Ortolano, L., K. Cushing and Contributing Authors (2000). *Grand Coulee Dam and the Columbia Basin Project*, USA. Case study report prepared as an input to the World Commission on Dams, Cape Town, *www.dams.org*

Patridge, Mark D. and Dan S. Rickman (1998). "Regional Computable General Equilibrium Modeling: A Survey and Critical Appraisal", *International Regional Science Review* 21(3).

Pyatt, G. (1994). "Modeling Commodity Balances in a Computable General Equilibrium Context", *Economic Systems Research* 6(2): 123-33

———. (1991a). "SAMs, the SNA and National Accounting Capabilities", *Review of Income and Wealth Series* 37(2): 177-98.

———. (1991b). "Fundamentals of Social Accounting", *Economic Systems Research* 3(3): 315-41.

———. (1988). "A SAM Approach to Modeling", *Journal of Policy Modeling* 10: 327-52

———. (1985). "Commodity Balances and National Accounts: A SAM Perspective", *Review of Income and Wealth*, Series 31(2): 155-69.

Pyatt, G.and J.L. Round (1985). "Accounting and Fixed-Price Multipliers in a Social Accounting Matrix Framework", in G. Pyatt and J.L. Round (eds.), *Social Accounting Matrices: A Basis for Planning*. Washington D.C.: The World Bank,

Robinson, S. (1989). Multisectoral models. in H. Chenery and T. N. Srinivasan. (eds) *Handbook of Development Economics*, Vol. II. Amsterdam: Elsevier Science Publishers.

Robinson, S., A. Cattaneo, and M. El-Said (2001). "Updating and Estimating a Social Accounting Matrix using Cross Entropy Methods", *Economic Systems Research* 13: 47.64.

Robinson, S., A. Yúnez-Naude, R. Hinojosa-Ojeda, J.D. Lewis, S. Devarajan (1999). "From Stylized to Applied Models: Building Multisector CGE Models for Policy Analysis", *North American Journal of Economics and Finance* 10: 5-38.

Robinson, S. and C. Gelhar (1995). "Land, Water and Agriculture in Egypt: The Economywide Impact of Policy Reform", *Trade and Macroeconomic Division Discussion Paper* No. 1. Washington D.C.: IFPRI.

Robinson, S. and M. El-Said (2000). "GAMS Code for Estimating a Social Accounting Matrix (SAM) using Cross Entropy Methods (CE)", *Trade and Macroeconomics Division Discussion Paper* No. 64. Washington D.C.: IFPRI.

Roe, T. and X. Diao (1994). *The Strategic Interdependence of a Shared Water Aquifer: A General Equilibrium Analysis*. Department of Agricultural and Applied Economics, University of Minnesota.

Rutherford, Thomas F. (1995). Extension of GAMS for Complementarity Problems Arising in Applied Economics. *Journal of Economic Development and Control* 19(8).

Scott, M.F.G. (1976). "Shadow Wages in Mauritius", in I. Little and M. Scott (eds.), *Using Shadow Prices*. London: Heinemann.

Shoven, J.B., and J. Whalley (1992). *Applying General Equilibrium*. New York: Cambridge University Press.

Squire, L., and H. van der Tak (1975). *Economic Analysis of Projects*. Baltimore: Johns Hopkins University Press.

Sinha, B. and R. Bhatia (1982). *Economic Appraisal of Irrigation Projects in India: A Case Study in Social Benefit Cost Analysis*. New Delhi: Agricole Publishing Academy.

Stone, R. (1985). "The Disaggregation of the Household Sector in the National Accounts", in G. Pyatt and J.I. Round (eds)., *Social Accounting Matrices: A Basis for Planning*. Washington D.C.: The World Bank,

Takayama, T and C.G. Judge (1964). "Equilibrium among Satially Seperated Markets: A Reformulation", *Econometrica* 32.

Taylor, L., E.L. Bacha, E.A. Cardoso and F.J. Lysy (1980). *Models of Growth and Distribution for Brazil*. New York: Oxford University Press.

USACAE (1933). Quoted in *USA Case Study: Final Scoping Report*. Available at *www.dams.org/report*

UNIDO (1972). *Guidelines for Project Evaluation*. New York.

Ward, W.A. and B.J. Deren (1991). *The Economics of Project Analysis: A Practitioner's Guide*. EDI Technical Material. Washington D.C.: The World Bank.

WCD (2000). *Dams and Development: A New Framework for Decision-Making*. the report of the World Commission on Dams. November 16. available at *www.dams.org/report*

Wigle, R.M. (1992). "Transportation Cost in Regional Medels of International Trade : An application to Canada-United States Trade", *Journal of Regional Science* 32(2).

World Bank (1996). "The World Bank's Experience with Large Dams—A Preliminary Review of Impacts". *Operations Evaluation Department Report* No. 15815. Washington D.C.: The World Bank.

Young, R.A. (2001). *Measuring Economic Benefits for Water Investments and Policies: Concepts, Approaches and Case Studies*. (Draft).

———. (1996). "Measuring Economic Benefits for Water Investments and Policies", *World Bank Technical Paper* No. 338. Washington D.C.: The World Bank.

Annex A-2.1

Details on Social Accounting Matrices and Computable General Equilibrium (CGE) Models

Social Accounting Matrices (SAM) and SAM-based Multiplier Analysis

The interest in income distribution and the flow of funds among "institutions" (household categories, firms, the government) led to the development of a more general social accounting framework than the 1968 System of National Accounts (SNA): the Social Accounting Matrix (SAM) (see Pyatt, 1994, 1991a; b; 1985, Pyatt and Round, 1985). Its structure can be translated into a set of linear equations with fixed coefficients, and used to compute fixed-price multipliers (Pyatt, 1988)

A Social Accounting Matrix (SAM) is a comprehensive, economywide data framework that represents the circular flow of income and expenditure in the economy of a nation or region. Each cell represents a payment from a column account to a recipient row account. Social Accounting Matrices synthesise two ideas in economics (Devarajan *et al.*, 1994). The first is the use of double-entry book-keeping that lies at the heart of I/O tables, where transactions between sectors representing inter-industry linkages appear in the accounts of both sectors involved. The SAM framework generalises the I/O idea to all transactions in the economy, so that any flow of money is recorded as a payment from a column economic agent to a row economic agent. The second idea descends from national income accounting, where income always equals expenditure, i.e., the budget constraint is respected. The SAM require balanced accounts (income = expenditure) for all actors in the economy. As a result, the SAM is square and the distribution of factor incomes to institutions is explicitly traced. These institutions generally comprise numerous household groups—distinguished on the basis of income, location, etc.—firms and the government. Table A.2.1 represents a standard SAM.[13]

13. Alternative SAM structures exist, but this is chosen because it is the one used as an input in the standard computable general equilibrium model developed by the International Food Policy Research Institute (Löfgren *et al.*, 2001).

Each cell of the SAM can be seen as a 'block' representing sets of transactions. As rows and columns total must be equal, expressions linking row and column elements with their totals can be used to form a system of linear equations that embodies market, behavioural and system relationships. The first have to do with the accounts relative to goods and factor markets. Behavioural relationships regard the budget constraints of economic agents institutions represented in the SAM. Finally, system relationships regard the capital and rest of the world accounts, where macroeconomic (internal and external) balances are represented.

A crucial distinction between SAM-based multiplier analysis and I/O models is that SAMs are able to account for the way in which initial asset distribution and factor endowments—and therefore the distribution of income among household categories and between them and the corporate sector—interact with the structure of production in determining final outcomes, particularly for welfare analysis. The key *raison d'être* for the SAM is that it embodies the fact that within the macro-economy there is a circular flow process and that what happens at one point on the circuit will have implications for experience at other junctures, so that one needs to be equally concerned with all the different aspects of technology and behaviour that together describe the circular income flow and the connections in the economy.

Pyatt (1991a) states that "[...]this is the full force of the general proposition that production structure and income distribution are inextricably interwoven, so that the one cannot be analysed properly without reference to the other[...]if one is interested in the distribution of income across socio-economic groups, i.e. in how different types of people are affected by policy change or exogenous events, then it is necessary to examine in corresponding detail what is happening in labour markets. It is not realistic to expect that poverty can be analysed without reference to unemployment." [p.194]

Assuming that some accounts are exogenous, the SAM can be expressed as a multi-sector, demand-driven model that considers the linkages between sector production, household incomes and expenditures, foreign trade and macroeconomic balances. The SAM can be translated into a set of accounting relationships. For endogenous accounts, total income (its row sum) is equal to the sum of products of A_n, the square matrix of (fixed) average propensities to consume, and y_n, the vector of endogenous incomes, plus the vector of total exogenous incomes x:

$$Y = A_n y_n + x$$

The SAM Multiplier Matrix M_a can thus be derived as follows:

$$Y = (I - A_n)^{-1} x = M_a x$$

Each cell of the multiplier matrix indicates the total (direct and indirect) income change in the endogenous row account income induced by an exogenous unit-income injection in the column account, thus capturing both the Leontief production linkages and the consumption linkages induced by changes in production activities through their effect on household incomes. Pyatt and Round (1985) show how the accounting multiplier matrix that can be derived from a SAM can be decomposed into three matrices,[14] one summarising the effects of transfers among endogenous institutions within the economy, one showing the cross effects of the multiplier process, i.e. how injections into one part of the system impacts other parts or 'open-loop effects', linked with the class of models that traces impacts of exogenous changes in income distribution on output & employment, with no allowance for effects in the reverse direction, and finally, one matrix recording the full circular effects of income injection going round the system and back to a point of origin in a series of repeated, dampening cycles.

It is important to note that these accounting matrices only capture the average effects of changes in injections into the economy on levels of endogenous incomes, but say nothing of behaviour at the margins. Multipliers, can, therefore be read as income impacts of exogenous changes only if one assumes all income elasticities to be equal to one. Even in the context of a fixed-price analysis, one may say something about this, but the above analysis needs to be integrated to consider

14. Pyatt and Round (1985) derive both a multiplicative and an additive form for their combination into the final multiplier matrix. The additive form is derived by Stone (1985).

Table A-2.1

Basic SAM Structure in the standard CGE model

	Expenditures								
Receipts	*Activities*	*Commodities*	*Factors*	*Households*	*Enterprises*	*Government*	*Savings-Investments*	*Rest of World (RoW)*	*TOTAL*
Activities		Marketed outputs	Home-consumed outputs						Activity Income (Gross Output)
Commodities	Intermediate Inputs	Transactions costs	Private Consumption			Government consumption	Investment	Exports	Demand
Factors	Value-added							Factor income from RoW	Factor income
Households			Factor income to household	Inter-household Transfers	Surrplus to households	Transfers to households from RoW		Transfers to households	Household income
Enterprises			Factor income to enterprises			Transfers to enterprises from RoW		Transfers to enterprises	Enterprise income
Government	Producer taxes, value-added tax	Sales taxes, tariffs, export taxes	Factor income to government, factor taxes	Transfers to government direct household	Surplus to government, direct enterprise taxes			Transfers to government from RoW	Government income
Savings-Investments				Household Savings	Enterprise Savings	Government Savings		Foreign Savings	Savings
Rest of the World (RoW)		Imports	Factor income to RoW		Surplus to RoW		Government transfers to RoW		Foreign exchange outflow
Total	Activity expenditures	Supply	Factor expenditures	Household expenditures	Enterprise expenditures	Government expenditures	Investment	Foreign exchange inflows	

how agents respond to changes at the margins. By complementing the SAM with information on income elasticities, the accounting multiplier matrix can be transformed into a fixed price multiplier matrix that accounts for the fact that increments of income will likely not be distributed in same proportions as the shares observed in the SAM.

Fixed-price multipliers, however, are still just one step ahead of simple accounting multipliers. Most of the limitations characterising I/O models—their demand driven nature, lack of price responsiveness, linear inter-sectoral interactions and factor use—also affect SAM-based multiplier analyses. However, they represent a step forward in that they close the loop between factor incomes, their distribution to different endogenous and exogenous economic agents, and their expenditure behaviour.

The next step requires the consideration of flexible prices and the substitution behaviours that are induced by them.

Figure A.2.1

The Circular Flow of Income in an Open Economy

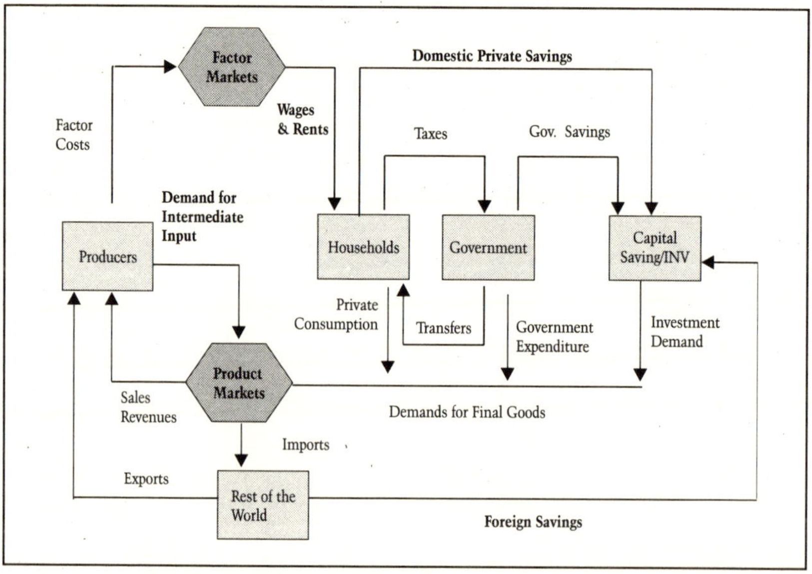

Computable General Equilibrium (CGE) Models

A CGE model is a system of simultaneous nonlinear equations that provide a complete and consistent picture of the 'circular flow' in an economy, capturing all market-based interactions amongst economic agents. Figure A.2.1 describes the circular flow of incomes for a simple open economy.[15] The behavioural assumptions determining agents' decisions stem from conventional microeconomic theory:[16] Producers buy factors of production from households and sell output on product markets to maximise profit, subject to technological constraints embodied in the production function; consumers maximise utility from consumption, subject to budget constraints.

Four features distinguish CGE models from their principal precursor, Leontief's input-output modeling tradition:

- Price endogeneity, as opposed to quantity adjustments, to reach an equilibrium.
- Price-responsive input and output substitutability—perfect or imperfect—through the use of nonlinear supply and demand equations (Robinson, 1989).
- The adoption of the so-called "Armington assumption" in terms of product tradability, whereby tradability is seen as a continuum where all goods are more or less tradable depending on a substitution elasticity (Armington, 1969), and therefore, the abandonment of the perfect dichotomy between traded and non-traded goods from traditional I/O models.
- Factor supply constraints, which generate output supply constraints.

Price endogeneity and general equilibrium interactions also set CGE models[17] apart from their other major antecedent: economy-wide

---

15. For each arrow, an arrow in the opposite direction indicating real flows exists, as CGE models require consistency between real and nominal flows.
16. This feature sets CGE models apart from macro econometric models. These generally consist of reduced form equations containing a few endogenous macro variables and failing to adequately represent agents' behaviour.
17. This is true of general equilibrium models in general, whether they are computable or not.

linear programming (LP) models.[18] In an LP, the weights in the objective function are given and a solution output vector is obtained by 'solving' the productive sphere of the economy alone.[19] The final demand vector is not linked to the factor incomes implicit in the solution, and there is, therefore, no feedback mechanism requiring an adjustment in prices. CGE models, on the other hand, include this basic feedback. Prices must endogenously adjust until the decisions made in the productive sphere are consistent with the final demand decisions made by households, whose incomes are endogenously determined by the equilibrium on factor markets. Equilibrium is reached through the interaction of independent decisions made by optimising economic agents, who react to the working of market-clearing mechanisms in product and factor markets.

Another relevant characteristic of CGE models is that neoclassical assumptions of perfect competition, perfectly functioning markets with flexible prices, and free products and factor mobility can be relaxed. Hence, they can be used to model special institutional features and market distortions—such as imperfectly competitive behaviour, quantity or price adjustment lags, and government interventions on prices or quantities. This enables researchers to assess the impacts of policies, projects or exogenous shocks in a second-best environment, and to compare their outcomes with the outcomes under first-best conditions. This feature is important from the point of view of our study in that the CGE models would enable us to compare the expected direct and indirect impacts of a dam project both in the context of the actual distortions present in the regional or national economy in which it operates, as well as in an undistorted context that is analogous to that assumed by the use of shadow prices for project evaluation.

Researchers of the International Food Policy Research Institute (Löfgren *et al.*, 2001), have developed a standard CGE model, which is

18. For a discussion of the relationship between linear programming and CGE models see Dervis *et al.*, 1982; Ginsburg and Waelbrock (1981); and Ginsburgh and Robinson (1984), which refers specifically to CGE models of developing countries.

19. Here it is assumed the LP optimises from the perspective of the firm by either maximising profit or minimising costs.

available on the internet at *www.ifpri.org* The paper includes an equation-by-equation description of the model and of its required database. The paper also discusses the implementation of the model in GAMS (the General Algebraic Modeling System[20]) and is accompanied by a self-extracting zip file, which includes the GAMS files for the model, sample databases, simulations, solution reports, and a SAM aggregation programme. Although the paper provides a standardised framework for analysis, the GAMS code is written so as to give the analyst considerable flexibility in model specification.[21] The following description closely follows that of Löfgren *et al.* (2001).

The standard model is based on the SAM presented above in Table A.2.1 In addition to the distinction between commodities and activities discussed in the previous section, the following characteristics of the SAM are worth mentioning.

1. Trade flows are explicitly associated with transactions (trade and transportation) costs, also referred to as marketing margins. For each commodity, the SAM records the costs associated with domestic, import, and export marketing. For domestic output, the marketing margin represents the cost of moving the commodity from the producer to the domestic demander. For imports, it represents the cost of moving the commodity from the border to the domestic demander, while for exports, it shows the cost of moving the commodity from the producer to the border. A services activity, called transportation, produces a commodity that, like other commodities, may be purchased for intermediate use by activities and final use by institutions. However, in addition, the transportation commodity receives payments from three special accounts, representing the transactions costs

20. See Brooke *et al.* (1988).

21. Brooke et al. (1998) is the basic reference on the GAMS software; it also includes a self-contained tutorial. The basics of GAMS-based CGE modeling are summarised in Robinson *et al.* (1999). Löfgren (2000a, 2000b) presents a set of exercises in CGE modeling with GAMS. Extensive treatments of CGE methods are found in Dervis *et al.* (1982); Robinson (1989); Shoven and Whalley (1992); Dixon, Parmenter, Powell and Wilcoxen (1992); and Ginsburgh and Keyzer (1997). References to and examples of CGE based analyses of food policy in developing countries are found in the section of the Trade and Macroeconomics Division on IFPRI's website (*www.ifpri.org*).

associated with domestic sales, imports, and exports. These special accounts are paid by the accounts for marketed agricultural and industrial commodities. This implies that the total value of each commodity includes these transactions costs. The standard model will also work with SAMs without this treatment of transaction costs.

2. The government is disaggregated into a core government account and different tax accounts, one for each tax type. This disaggregation is often necessary to avoid ambiguities in the economic interpretation of some payments. In any given application, the SAM may exclude any (or all) of the individual tax accounts. In the SAM, payments between the government and other domestic institutions are transfers
3. Depending on data availability, it is preferable to have one or more accounts for enterprises when these have tax obligations and a savings behaviour that are independent of the household sector. It is important to disaggregate the enterprise sector in a manner that captures differences across enterprises in terms of tax rates, savings rates, and the shares of retained earnings that are received by different household types. For example, it may be appropriate to disaggregate enterprises into non-agricultural (with earnings from non-agricultural capital), small-scale agricultural (earnings from land and capital controlled by small farmers), and large-scale agricultural (earned from the land and capital of large farmers). Note however, that only one household account is required, while enterprise accounts are not necessary.
4. The SAM distinguishes between home consumption of activity output, and household marketed consumption of commodities. Home consumption appears in the SAM household payments to activities and is valued at producer prices. Household consumption of marketed commodities appears as payments from household to commodity accounts, and their values include marketing margins and commodity taxes. The standard model also accepts a SAM without (explicit) home consumption.

The model explains all the payments that are recorded in the SAM, and therefore follows the SAM disaggregation of factors, activities, commodities, and institutions. It is written as a set of simultaneous equations, many of which are non-linear, with no objective function. The equations define the behaviour of the different actors and include a set of constraints that have to be satisfied by the system as a whole but which are not necessarily considered by any individual actor. These constraints include market balances (for factors and commodities) and macroeconomic aggregates (balances for savings-investment, the government, and the current-account of the rest of the world). These systemic constraints can take a number of different forms and are important in characterising the nature of the model. The section on macro closures discusses the rules for clearing the macroeconomic balances, i.e., how equilibrium is achieved in the balances for the government, the rest of the world, and the savings-investment account (where institutional savings are aggregated and allocated to domestic investment).

Activities, Production, and Factor Markets

Each producer (activity) is assumed to maximise profits, defined as the difference between revenue earned and the cost of factors and intermediate inputs. Profits are maximised subject to a production technology, which has a nested structure. At the top level, the technology is specified by a constant elasticity of substitution (CES) or, alternatively, a Leontief function of value-added and aggregate intermediate input. The Leontief alternative is the default. The CES alternative may be preferable in particular sectors if available technique permits substitution between value-added and intermediate inputs. Value-added is itself a CES function of primary factors, while the aggregate intermediate input is a Leontief function of intermediate inputs.

Each activity produces one or more commodities according to fixed yield coefficients. Revenues are defined by the level of output, yields, and commodity prices at the producer level. As part of its profit-maximising decision, each activity uses factors up to the point where the marginal revenue product of each factor is equal to its wage. Wages may differ across activities, not only when the market is segmented (i.e., when factors are either sector- or region-specific) but also for mobile factors.

In the latter case, the model incorporates sectoral wage discrepancies that stem from exogenous causes (e.g., status, comfort, or health risks).

The user can choose between alternative factor market closures. (1) The default closure (Kaldorian[22]). Quantity supplied of each factor is fixed at the observed level, and an economy-wide wage variable adjusts to assure that the sum of demands from all activities equal the quantity supplied. Each activity pays an activity-specific wage that is the product of the economy-wide wage and an activity-specific wage distortion term, which are fixed. (2) The Keynesian closure. One factor is unemployed and its real wage is fixed. Compared to the default closure, the only change is that the economy-wide wage variable is exogenous, while the supply variable varies endogenously. Each activity is free to hire any desired quantity at its fixed, activity-specific wage (which, implicitly, is indexed to the model numéraire). In this setting, the supply variable is superfluous; it merely records the total quantity demanded. (3) Segmented factor markets. The factor market is segmented and each activity is forced to hire the observed, base-year quantity—the factor is activity-specific. This closure may be preferred in short-run analysis and/or when there are significant quality differences between the units of a factor that are used in different activities. The quantities of activity-specific factor demands and the economy-wide wage are fixed, while the activity-specific wage terms and the supply variables vary.

Institutions

Institutions are represented by households, enterprises, the government, and the rest of the world. The households (disaggregated as in the SAM) receive income from the factors of production directly, or indirectly via the enterprises, and transfers from other institutions. The households use their income to pay direct taxes, save, consume, and make transfers to other institutions. In the basic model version, direct taxes and transfers to other domestic institutions are defined as fixed shares of household income whereas the savings share is flexible for selected households. The treatment of direct tax and savings shares is related to the choice of closure rule for the government and savings-

22. Bandara (1991)

investment balances. The income net of taxes, savings, and transfers to other institutions is spent on consumption.

Household consumption covers marketed commodities, purchased at market prices that include commodity taxes and transactions costs, and home commodities, which are valued at activity-specific producer prices.[23] Household consumption is allocated across different commodities according to Linear Expenditure System (LES) demand functions. Other demand functions can be used (CES, AIDS, etc.)

Enterprise factor income reflects ownership of capital and/or land, and may receive transfers from other institutions. They use their incomes for direct taxes, savings, and transfers to other institutions, but do not consume. Apart from this, payments to and from firms are modeled in the same way as the same payments to and from households.

The government collects taxes and receives transfers from other institutions. In the basic model version, all taxes are at fixed *ad valorem* rates. The government uses this income to purchase commodities for its consumption and for CPI-indexed transfers to other institutions. In the basic model version, government consumption is fixed in real (quantity) terms whereas government transfers to domestic institutions (households and enterprises) are CPI-indexed. Government savings (the difference between government income and spending) is a flexible residual.

Finally, transfer payments between the rest of the world and domestic institutions and factors are all fixed in foreign currency. Commodity trade with the rest of the world is discussed in the following section. Foreign savings (or the current account deficit) is the difference between foreign currency spending and receipts.

Commodity Markets

Domestic output may be sold in the market or consumed at home. For marketed output, aggregated domestic output for each commodity is computed as the composite output of different activities that produce

23. In the standard SAM, home consumption is only disaggregated by activity and household, not by commodity, activity, and household. When households consume out of activities that produce multiple outputs, non-SAM data is needed to allocate home consumption across the commodities produced by multiple-output activities.

the commodity. These outputs are imperfectly substitutable, for example as a result of differences in timing, quality, and location between different activities. A Constant-Elasticity-of-Substitution (CES) is used as aggregation function. The demand for the output of each activity is derived from the problem of minimising the cost of supplying a given quantity of aggregated output subject to this CES function. Activity-specific commodity prices serve the role of clearing the implicit market for each disaggregated commodity.

The separation of activities and commodities into two accounts also provides the data needed to model imports and exports as perfect or imperfect substitutes vis-à-vis domestic production. At the disaggregated commodity level, these assumptions allow for two-way trade, which commonly is observed even at very fine levels of disaggregation

The Armington assumption treats imports (M) and demand for domestically produced goods (DC), having the same sectoral (commodity) classification, as differentiated goods whose demand is characterised by a specified elasticity of substitution. On the production side, this specification introduces an elasticity of transformation between goods supplied to domestic markets (DA) and those for the export market (E) by each sector (activity). This differentiation on both the export and import side introduces a degree of autonomy of the domestic price system from changes in world prices of imports and exports and from exogenous or policy-induced changes in the exchange rate.

Under the small country assumption, i.e., the country's imports have an infinitely elastic world supply and its exports have an infinitely elastic world demand, world prices of imports pwm and of exports PWE are exogenously determined. The domestic prices of imported and exported products are given by:[24]

$$PM_i = pwmi\ (1 + tm\ i)\ EXR$$

$$PE_i = pwei\ (1 - te\ i)\ EXR$$

respectively, where EXR is the exchange rate (in domestic currency per unit of foreign currency), and tm and te are the implicit tariff and export

24. The notation that follows is from Bautista *et al.* (1999).

tax rates that account for legal tariffs and export taxes, as well as any quantitative trade restrictions and direct price controls that affect the disparity between the domestic and border prices of traded goods.

When this simplifying assumption is relaxed, consumer face commodities that are composite goods Q, whose demand is a constant elasticity of substitution (CES) aggregation function of M and D, with a substitution elasticity s. Consumers maximise utility, which in the model is the same as maximising consumption Q, so that the desired ratio between M and D is a function of their relative prices. In multi-sector models utility maximisation takes place at two levels: (i) to allocate expenditure among goods, and (ii) to decide optimal import ratios. Similarly, producers provide a composite commodity X that is a constant elasticity transformation (CET) function of E and D and maximise profits so that their desired ratio is a function of their relative prices.

$$(M\ i\ /\ DC\ i) = CES^{\star}(PM\ i/PDC\ i)$$

$$(E\ i\ /\ DA\ i) = CET^{\star}(PE\ i/PDA\ i)$$

Where CES* and CET* refer to the first-order conditions for utility maximisation and profit maximisation.

Sectoral composite good prices are the weighted averages of the domestic prices of their component products:

$$PQ\ i = (PDC\ i \bullet DC\ i + PM\ i \bullet M\ i)\ /\ Q\ i = CES\ (PDC\ i\ ,\ PM\ i) \text{ and}$$

$$PX\ i = (PDA\ i \bullet DA\ i + PE\ i \bullet E\ i)\ /\ X\ i = CET\ (PDA\ i\ ,\ PE\ i) \text{ and}$$

where the CES and CET functions refer to cost functions relating the composite prices to their component prices. They reflect the first-order conditions described above.

This model specification determines the role of the exchange rate in achieving equilibrium, as it turns it into a well-defined relative price. The model specifies a functional relationship between the balance of trade and the real exchange rate—i.e., the relative price of tradable and semi-tradables—which is one of the crucial mechanisms that drive how external shocks and policies affect the real side of the economy in the model.

Note that the model code makes it possible to treat selected imports as separate, 'non-comparable' commodities that have no corresponding domestically produced good. In the commodity rows, such import commodities receive payments from one or more domestic users. In the columns, these payments would be passed on to the accounts for the rest of the world, import marketing margins, and relevant taxes. The columns for this category of imports do not have any payments to domestic activities.

Macro Closures

The model includes three macroeconomic balances: (i) the (current) government balance, (ii) the current account of the balance of payments, which includes the trade balance, and (iii) the savings-investment balance. In the GAMS code, the user has the option to choose among a relatively large number of pre-programmed alternative closure rules for these balances.

For the government balance, the default closure sets government savings as a flexible residual, while all tax rates are fixed. Under two alternative government closures, the direct tax rates of domestic institutions (households and enterprises) are adjusted endogenously to generate a fixed level of government savings. The first of these alternative closures, adjusts endogenously the base-year direct tax rates of selected domestic non-government institutions by the same number of percentage points. In the second, the rates of selected institutions are multiplied by a flexible scalar. All three government closures fix government consumption, either in real terms or as a share of nominal absorption (depending on the treatment of the savings-investment balance). The standard model does not specify a closure where government savings and direct tax rates are both fixed and government consumption is the adjusting variable, but this is indeed possible and can be programmed.

The external balance is expressed in foreign currency. The default closure has a flexible endogenous real exchange rate, while foreign savings (the current account deficit) is fixed. Given that all other items in the external balance (transfers between the rest of the world and domestic institutions) are fixed, the trade balance is also fixed. If, *ceteris paribus*, foreign savings are below the exogenous level, a depreciation of the real exchange rate would correct this situation by simultaneously (i) reducing spending on imports (a fall in import quantities at fixed world prices); and (ii) increasing earnings from exports (an increase in export quantities at fixed world prices). An alternative closure imposes a fixed real exchange rate (indexed to the model numéraire), while foreign savings (and the trade balance) are flexible.[25]

For the savings-investment balance, closures are either investment-driven (the value of savings adjusts) or savings-driven (the value of investment adjusts).[26] The default closure is investment-driven. Real investment quantities are fixed and savings adjust to equal the cost of the investment bundle. In the default closure, the base-year savings rates of selected non-government institutions are adjusted by the same number of percentage points. An investment-driven alternative differs from the default in that the rates of selected institutions are multiplied by a scalar. Again, government consumption and savings are kept constant, so that the brunt of the adjustment falls on household absorption. Another alternative is savings-driven. All non-government savings rates are fixed. The quantity of each commodity in the investment bundle is multiplied by a flexible scalar so as to assure that the investment cost will be equal to the savings value.

Finally, two "balanced" closures are possible, which may be viewed as variants of investment-driven closures that also impose an adjustment rule for government consumption. Under these, adjustments in absorption are spread across all of its components. The nominal absorption shares of investment and government consumption are fixed at base levels, while in previous closures, government consumption is fixed in real terms. Given this specification, the residual share for household consumption is also fixed. For the first balanced closure, the savings rates of selected institutions are adjusted

25. See Devarajan *et al*. (1994).

26. Bandara (1991) terms these the Johansen closure and the Classical closure, respectively.

by an equal number of percentage points. For the second balanced closure, the savings rates of selected institutions are scaled so as to generate enough savings to finance investment. The balanced closures are compatible with any combination of the preprogrammed closures for the government and the rest of the world.

The appropriate choice between the different macro closures depends on the context of the analysis. For a single-period model, a closure combining fixed foreign savings, fixed real investment, and fixed real government consumption may be preferable for simulations that explore the equilibrium welfare changes of alternative policies. Such a closure avoids the misleading welfare effects that appear when foreign savings and real investment change in simulations with a single-period model—*ceteris paribus*, for the simulated period, increases in foreign savings and decreases in investment raise household welfare. This result is misleading since the analysis does not capture welfare losses in later periods that arise from a larger foreign debt and a smaller capital stock. With regard to government consumption, the model does not capture its direct and indirect welfare contributions; to avoid misleading results, it is also preferable to keep this variable fixed. On the other hand, if the analysis aims at capturing the likely welfare effects of an exogenous shock or policy change in a given (historical, current, or future) setting, perhaps in order to explore the role for complementary policies, it may be preferable to impose a closure that more closely mimics the real world, often involving simultaneous adjustments in the three components of absorption. Under these circumstances, a macro scenario that incorporates a balanced closure is a useful option.

The second option is likely to be more relevant for the analysis of welfare impacts of large dam projects. However, it may be informative to explore its impacts under a set of alternative macro closures, as this can provide important insights into the trade-offs that are associated with alternative macroeconomic adjustment patterns.

Recent Advancements in CGE Modeling: A Brief Review of those Relevant for our Analysis

CGE modeling is advancing in a number of new directions. Among them, the development of 'mixed complementarity' CGE (CGE-MC) models has proven especially fertile for the analysis of water resources use. CGE-MC models can be seen as a rediscovery of the mathematical programming roots of CGE modeling, in an attempt to overcome the limitations imposed on them by the consistency model nature of the input-output structure lying at the core of standard CGE models. This aspect imposes on them a property of 'conservation of flow', whereby output, input use, exports and imports cannot be discontinued if they exist in the initial data set representing the baseline, and cannot "emerge" if they are not present in the baseline. In recent formulations, mathematical programming techniques have significantly contributed to overcoming these limitations, injecting flexibility into this structure by introducing inequality constraints.

Following the Kuhn-Tucker formulation of an optimisation problem, CGE-MC models combine equalities and inequalities. Each inequality is linked to a bounded variable in a complementary slackness condition.[27] This model specification permits the incorporation of the following features:

- The endogenous determination of factor and resource market regimes. Resource underemployment, associated with a zero scarcity wage, is allowed by introducing inequalities in the specification of the equilibrium on factor and resource markets. This is done by substituting activity analysis (or "Leontief", or "fixed coefficient") production functions for the smoothly substitutable neoclassical production or cost functions whose flexible wages, rents, and prices produce full employment market equilibria.
- A more realistic specification of technology and supply. Where relevant, output can be produced using more than one alternative technique (i.e., activities) and the production regime is endogenous, as some of the techniques can be de-activated, so that some of the activities can see their output go to zero. This type of change in the structure of the economy *verses* the initial

27. See Ginsburgh and Waelbroek, (1981); Rutherford, (1995); Löfgren and Robinson, (1997).

"snapshot" embodied in the SAM is not possible using standard CGE models.

- The capacity to introduce upper and lower bounds on relevant variables, permitting the simulation of policies that set targets or introduce some form of rationing.

Two general approaches exist for incorporating water resources in a CGE model, each best suited to study a different set of issues. The first approach considers water as a factor of production found in nature. Its existing examples follow the CGE-MC specification, extending standard CGE models by nesting into their structure an activity analysis representation of water and land use by agricultural sectors. Land and water are available in fixed amounts. Their demands are combined according to fixed coefficients into the demand for a water/land aggregate factor, which is in turn combined with labour and capital to generate value-added. The price of the land/water aggregate is a linear combination of land and water prices, which at equilibrium equal the shadow prices on the inequality constraints on water and land availability.

Models that follow this approach have been used to study issues such as increasing water scarcity and the consequent changes in the scarcity value of water, uneven regional water distribution, and water allocation based on its scarcity value. Berck *et al.* (1991) present a CGE model of agricultural water use in the San Joaquin Valley, California. Robinson and Gelhar (1995) model arable land and water scarcity in Egypt to analyse the consequences of creating a water market and of introducing charges on agricultural water use. Mukherjee (1996) presents a single region CGE model for the Olifants river area in South Africa, to study inter-sectoral water allocation and its regional economic impacts. In these models, water services are not considered as a produced commodity, and this approach is therefore, not ideally suited to study cost recovery issues for water service provision.

A second approach is the one adopted by Decaluwé *et al.* (1997),[28] which consider water as a commodity produced by a regulated monopolist, sold on regulated markets at regulated prices, and used

28. A similar model, applied to a shared aquifer situation is Roe and Diao, (1994).

both as an intermediate input and as a final consumption good. The authors build a two-region model that specifies two different water production techniques whose parameters vary by region.[29] To reach an equilibrium, the model adjusts the output of the water production activities in the two regions, given water prices. This approach can be used to study the impacts of water pricing reforms to achieve cost recovery, but not to analyse scarcity issues and the value of water as a natural resource.

The second innovative path that CGE models are following involves the creation of multi-region CGE models, where space is explicitly considered and the regions of interest are at a sub-national level. Multi-region models should be chosen over standard national CGE models when the regional impacts of changes in policies and exogenous shocks are expected to be significantly different from the regional average. On the other hand, even if the main focus of the analysis is a single sub-national region—such as a river basin—multi-region models are to be preferred to single region models because the latter do not capture relevant inter-regional and regional-national feedbacks.

Despite the widespread use of regionally disaggregated (within country or multi-country) input-output and CGE models,[30] only a few recent examples are really able to deal with geographical space.[31] These models explicitly represent the existence of physical distances between regions and the restrictions they impose on trade and other flows, focusing primarily on transportation costs. Löfgren and Robinson (1999) represent an economy as a spatial [node-link] network and combine the explicit treatment of space with the CGE-MC formulation discussed above to enable endogenous trade regime shifts, linking this model to the spatial equilibrium tradition.[32] These advancements can greatly improve the application of CGE modeling tools to the analysis of water resource use and management, as they make it possible to model the

29. Note that the spatial dimension is not meaningfully incorporated in this model as (i) there are no water exchanges between the two regions; and (ii) physical distances do not give rise to any transportation costs.

30. For a review, see Partridge and Rickman, (1998).

31. Buckley, (1992); Wigle, (1992); Elbers, (1995); Löfgren and Robinson, (1999).

32. The seminal paper is Takayama and Judge, (1964).

consequences of the mobile nature of water, such as the changes in the quality of the resource as it moves through a basin or the transaction costs associated with transfers between users located in different regions.

Data Requirements: How to Build a SAM

To build a SAM, one should start from building a balanced aggregate macro-SAM that provides the key control totals that can then be used to help balance the disaggregated micro-SAM needed for the analysis. The basis for the aggregated macro-SAM are data from the current account of the balance of payments, the national accounts, the government budget and wage statistics. The more disaggregated micro-SAM requires the existence of an input-output table for the region of interest—or the use of the I/O table for a region whose structure of inter-industry linkages is assumed to be sufficiently close to that of the region in question. This information is then integrated with a number of other scattered data sources. There exist cases where SAMs have been built in the absence of I/O tables, a recent example exists for Mozambique (Arndt *et al.*, 1998)

IFPRI's Trade and Macroeconomics Division's web page (*www.ifpri.org*—click on Research, then Research Divisions, then Trade and Macroeconomics, then Divisional Output, then Discussion Papers) presents a number of studies, downloadable in Acrobat format, that describe data sources and methods that were followed to build SAMs in a number of countries, often with very sparse data. Modelers for the various case studies can refer to this literature, which is presented in a separate section of the references below.

In addition to a balanced micro-SAM at the appropriate disaggregation level, a CGE model requires the use of estimated values or "guesstimates" of a number of price elasticities, income elasticities and elasticities of substitution and of transformation, depending on the functional forms chosen for production and input demand functions, the aggregation of domestically produced and imported commodities, the aggregation of products for domestic and export markets, and commodity demand functions.

The GAMS programme found as part of the standard CGE model at *www.ifpri.com* checks that the SAM that is entered is balanced, i.e., that for each account, the row and column totals are equal. If the absolute value of the sum of account imbalances exceeds a cutoff point, an optimisation programme is used to estimate a balanced SAM. The programme, which minimises the entropy distance of the cells of the estimated SAM from those of the initial SAM subject to the constraint that row and column totals are equal, is primarily intended to remove rounding errors.

For SAM estimation in GAMS in a setting with substantial imbalances in raw data (not only rounding errors), see Robinson and El-Said (2000) and Robinson, Cattaneo, and El-Said (2001). Their approach applies information theory to estimating a system of nonlinear simultaneous equations. A number of characteristics make it appropriate for use in this setting.

First, it imposes all general equilibrium constraints. Second, it permits incorporation of prior information on parameter values. Third, it can be applied in sparse data situations. Finally, it supplies measures of the capacity of the model to reproduce the historical record and the statistical significance of parameter estimates. The method is applied to estimating a CGE model of Mozambique. These sources emphasise that any good source of data should be brought to bear in defining the constraints that are used to balance the SAM.

3 Comparative Analysis of Results of the Four Case Studies

Ramesh Bhatia, R.P.S. Malik, and Rita Cestti

The Four Case Studies: An Overview

As discussed earlier, four case studies were undertaken as part of this study with the common objective of estimating the indirect economic impacts of dams. While the broad methodological framework has been discussed in Chapter II, the finer details of the methodology followed in each case study and the estimated results obtained are discussed in detail in Chapters VI to IX. This Chapter gives an overview of the comparative results of the four case studies. However, before we proceed to present comparative results on value added multipliers and the income distribution aspects of these dams, we give below a brief description of each of the four dams.

Bhakra Multipurpose Dam System, India

The Bhakra dam system in the northern part of India has contributed significantly to increases in the output of agricultural commodities and electricity over the last 45 years or so. Additional gross irrigated area has been of the order of 7.1 million hectares per year. The total foodgrain production in the Bhakra system command area during the year 1996-97 was of the order of 27 million tonnes, an additional output of about 24 million tonnes compared to the food output in the early 1960s. The hydro power stations installed in the Bhakra system have a combined generating capacity of 2880 MW, which currently generate about 14000 million units (kWh) of electricity in a year. These increases have inevitably generated downstream growth in many other sectors of the regional economy as well as in other parts of the country.

Substantial marketed surplus of foodgrains in Punjab and Haryana have provided foodgrains for the urban poor all over India at relatively low food prices. Such surplus foodgrains have been used for the 'Food for Work' programme in many states thus helping the rural poor, especially during drought conditions, and generating multiplier effects. Hundreds of thousands of migratory workers from underdeveloped areas of Bihar and Uttar Pradesh have sent large amounts of remittances creating further multiplier effects in those regions. The availability of irrigation water and the consequent growth in rural and urban areas that followed has helped in a very significant way in bringing down poverty in the two states of Punjab and Haryana. As already mentioned, the benefits of this economic development in the region have been shared both by rural as well as urban areas. As a result this growth scenario has led to a steep decline in poverty in the rural areas of both the States—this is despite the fact that a very large number of labour from poor regions of the country having migrated to this region. The poverty in the two States is much lower in comparison to both all-India figures as also in comparison to other major States of the country. In rural areas, the percentage of people below the poverty line during 1999-2000 were only 6.35 percentage in Punjab and 8.27 per cent in Haryana as compared with 27 per cent in India. In urban areas, the per cent of poor people below the poverty line were only 5.7 in the Punjab and 10 per cent in Haryana compared to 23.6 in the country as a whole. The Bhakra dam system has played a very significant role in the last about 45 years in bringing about significant reductions in poverty in the Punjab and Haryana states.

The case study for the Bhakra dam system presents an analytical framework and some preliminary estimates of regional value added multipliers arising from inter-industry linkages of production and consumption-induced effects. A fixed-price SAM (Social Accounting Matrix) based multiplier model for the state of Punjab for the year 1979-80 has been used to estimate scenarios under 'With Project' and 'Without Project' situations. The aspects of the dam that have been analysed include: changes in the area irrigated; changes in the supplies of electric power; changes in yields and production technology (primarily changes in fertiliser use) that would have been likely in its absence. This analysis captures the main effects of the dam in a typical year (1979-80) during its lifetime, approximately 20 years after the construction was completed.

Bunga Check Dams, India

Two check dams were built in Village Bunga, located in the Shivalik Hills regions in northern India. The first of these dams was built in 1984 and the second in 1986. The second dam, however, could not deliver water due to some bureaucratic hurdles in the laying of pipelines for distribution of water. The second dam actually started supplying water in the year 1996. The two structures combined together are presently supplying irrigation water to 276 ha of village cultivated land. The Bunga case study attempts an estimation of multiplier effects of these check dams. It is useful to estimate multiplier effects of check dams as compared to those of large dams, even though check dams may not be substitutes of large dams in many situations.

The major output of the check dams has been the availability of water for irrigation. However, since the requirements for water for fully irrigated agriculture are much more than the water, this dam can store and supply, this dam has been able to fulfill only a part of the irrigation water requirements of the village. The other important outputs of the dams have been—almost complete elimination of floods that used to inundate these villages due to heavy run off of rain water during the rainy season; reduction in soil erosion from the barren hill slopes, and improved availability of drinking water in the villages for human beings as also for the livestock.

To capture the complex inter-linkages among village production activities, village institutions and the outside world, a village level Social Accounting Matrix (SAM) was developed. In order to unravel the impact of increases in agricultural commodities and milk production from all the other autonomous sources of growth in the economy, it was necessary to construct a picture of the regional economy as it would have been in 2002 had the Bunga check dams not been constructed. Multiplier values have been estimated by estimating differences in regional value added under 'With Project' situation and 'Without Project' situation and comparing these with differences in direct value added under the two situations. The income distribution impacts of the dams have been analysed for the households having land (various size groups) as compared to landless households.

The High Dam at Aswan, Egypt

Agriculture is a major source of income and economic activity in Egypt. According to 1992 government estimates, 57 per cent of the country's 40 million population was involved in agriculture, accounting for 31 per cent of total GNP and 50 cent of all exports. The building of the Aswan Dam has spurred urban growth, but the water shortage problem remains. The city of Aswan is a good illustration of the impact of urbanisation on the water source. As a result of the dam, the population of Aswan City grew from 63,000 in 1960 to 128,000 in 1966, and to 144,000 in 1976. In the period from 1957 to 1961, there was an employment surge from 2.9 thousand to 5.7 thousand and the value of production also quadrupled from L.E. 3.5 million to L.E. 14.0 million.

For the Aswan High Dam, a static CGE model has been used to simulate Egypt's economy with and without the High Dam at Aswan, drawing on an existing model and a review of the literature of cost-benefit analyses of the dam. For purposes of comparison, the impacts are also simulated with the model run in a fixed-price multiplier mode (replicating a SAM multiplier analysis). The database relies on a Social Accounting Matrix for 1997. The effects of the dam have been analysed comparing the current situation of reliable, constant water flow out of the dam with hydrological data regarding the natural flows of the Nile in its absence. Under these two scenarios, the model simulates: changes in the supplies of irrigated land and water; changes in yields and production technology; and changes in the supplies of electric power. The benefits that can be attributed to the dam have been compared with the real costs associated with the investment. This analysis captures the main effects, both costs and benefits, of the dam during a typical year during its lifetime. The indicators, capturing the effects, include production, trade, as well as disaggregated household incomes and their distribution.

Sobradinho Dam and the Cascade of Reservoirs in the Sub-Médio São Francisco River, Brazil

The São Francisco river is the largest river of Brazil's Northeast region, indeed the longest of all Brazilian rivers. It is also one of the few permanent rivers in a region characterised by spatially and temporally irregular rainfall patterns, high average temperatures, strong exposure to sunlight, and high evaporation rates. Most of its rivers are intermittent unless regulated, flowing only during the rainy season (Jan./Feb. to June). Water scarcity conditions are prevalent and have recently extended to wet months as well. Cyclical droughts have been occurring with increasing frequency, recurring every five years or less and, in particularly critical situations, persisting over more than one year.

The analysis in this case study focuses on Sobradinho dam and the cascade of reservoirs downstream of it. The construction of the Sobradinho dam in 1973-1979, was one of the most important factors in transforming the region's economy, society, and landscape. At the time of its construction, a popular catchphrase was that the sertão—the arid interior of Brazil's Northeast—would turn into a sea. The Sobradinho lake has over 4,000 Km^2 of surface areas and a water storage capacity of 34 billion m^3, alimenting a hydroelectric power plant with a capacity of 1 million kW. The dam has contributed not only to stabilising downstream flows—enabling the construction of a hydropower complex with an installed capacity of almost 10 GW—but also in providing water for large irrigation projects that are transforming agricultural production in the region. And yet, some of socio economic sources of vulnerability of the region appear to persist.

This case study analyses the direct and indirect economic impacts of damming the Sub-Médio São Francisco river, in an attempt to understand whether these impacts have indeed, as some maintain, fallen short of expectations and, if so, whether factors outside the scope of dam design and water use decisions are responsible. In particular, the evolution of the policy environment surrounding the dams and their affected sector have been discussed. The study also highlights some of the success stories that have emerged in the region, again trying to reveal both their multiplier impacts and the lessons to be learned from them.

The analysis is based primarily on a 1992 input-output (I/O) table for the Northeast region. An ongoing World Bank study on irrigation development impacts in Brazil's Northeast macro-region, and a concluded set of studies on the same subject supported by the Inter-American Development Bank were additional assets, both in terms of data collection and to support the design of scenarios and interpretation of their results. Multiplier and distribution analysis are based on a detailed analysis of poverty reduction impacts of agriculture in the region.

Table 3.1 summarises the methodological aspects of the four case studies while some of the main features of the four case studies have been summarised in Table 3.2.

The Four Case Studies: Some Similarities and Differences

As would be observed, even though all the four case studies undertaken as part of this study have the common objective of estimating the indirect economic impact of dams, the methodology employed and the results obtained in each case study are not strictly comparable. The

Table 3.1

Case Studies and Methodologies Followed for Estimation of Multipliers Effects

Case Study	*Size of Dam*	*Country/Region*	*Methodology*
Bhakra Dam System	Large	Northern India	Social Accounting Matrix (SAM) Based multiplier model with embedded Input-Output Model based on a State-level Social Accounting Matrix (SAM)
Bunga Check dams	Small	Northern India	Social Accounting Matrix (SAM) based multiplier model for the Bunga village
The High Dam at Aswan	Large	Egypt	Computable General Equilibrium (CGE) Model coupled with Mathematical Programming Model.
Sobradinho Dam and the Cascade of Reservoirs on the Sub-Médio São Francisco River	Large	Brazil Northeast	Semi Input-Output (S/I-O) Model based on a Regional Input-Output Table

Table 3.2

Salient Features of Selected Case Studies

Region	*Country*	*Height of Dam/ Type of Dam/ Size of Reservoir*	*Area Irrigated/ Production of foodgrains per year*	*Installed Capacity and Annual Generation*	*Major Outputs*
A. Large Dams					
Bhakra Dam System1/	India	225 meters Concrete and Earth-filled 18 BCM	10.3 Million Ha 27 Million Tonnes	2800 MW 14 Bn kWh	Irrigation, Drinking Water for several rural and urban areas in the command as well as also for Delhi Hydropower shared by at least Five States
Aswan High Dam	Egypt	111 meters 164 BCM Rock-filled	2.65 Million Ha	2100 MW 10 Bn kWh	Irrigation and Hydropower
Sobradinho dam (S) and reservoirs along the Sub-Médio and Baixo São Francisco river basin:	Brazil	(S): 41m x 8.5 km 4214 km^2 34100 MCM	São Francisco Basin: 330,000 ha (122,000 ha public irrigation schemes)	(S): 1050 MW All 1992: 6800 MW All to date: 9800 MW	Water and power for cities and towns in numerous Brazilian states
B. Small Check Dams					
Bunga Villages	India	2 check dams	276 Ha		Irrigation water, drinking water for human beings and livestock

***Notes*:** (1) Bhakra dam system includes Bhakra dam, Pong dam and Nangal dam

(2) The set of cascading reservoirs includes Itaparica (Luis Gonzaga) dam, the Paulo Afonso I-IV complex, Moxotó (Apollônio Sales) dam and Xingó dam. The latter is not part of the analysis, however, as it became operational only in 1995.

choice of methodology and aspects of analysis in each case study have in large part been influenced by the nature of data available from some of the related ongoing studies in the respective study regions. For example, due to non-availability of data relating to aspects of income distribution, the income distribution analysis has not been undertaken in the Brazil case, while data availability in the case of the other three case studies have permitted such an analysis.

An ongoing study in Egypt, has enabled the use of a fully operational CGE model for the analysis of the impact of the dam. For case studies of the Bhakra dam and check dams in Bunga, no previous studies using SAM-based models or CGE models are available. Based on such considerations, the Value-Added multipliers in the case of Sobradinho Dam (and the set of cascading reservoirs) in Brazil have been estimated using a Semi-Input-Output model; in the case of Bhakra dam and Bunga check dams these have been estimated from SAM-based fixed-price multiplier models; while the Aswan High Dam case study employs a CGE model.

All case studies compare 'With' dam and 'Without' dam scenarios. In all cases, the 'Without' dam scenarios have not been built as full-fledged counterfactuals of the 'With' dam case. The exogenous modification of a few parameters or initial variable levels does not fully capture the alternative configuration that the region's economy would have had, had an alternative development path been chosen. The case studies start from a 'typical' operational year in the mature life of the selected dams and simulates the impacts of removing the dam and its direct output from that context—a context that was inescapably determined by the very construction of the dam years before. Great care has therefore, been taken in the attempt to both design meaningful 'Without' dam scenarios and interpret their results, to avoid drawing conclusions that do not consider these serious limitations.

In all case studies concerning large dams, the river basin where these are located is a focus of the analysis. As it was not always possible, nor always appropriate,[1] to build databases of the river basin

1. When the impacts of a dam go beyond the river basin where this is located.

economy, the studies considered the river basin level to build scenarios, interpret results, and/or extend the analysis beyond the modeling activities to consider impacts that may not be assessed based on the model. For example, for the Bhakra dam system, the entire river basin is considered even though multiplier estimates using the SAM model are confined to its portion located in the Punjab state.

In addition to estimating multipliers, discussing the factors affecting their magnitude, and assessing income distribution impacts of each dam, wherever possible, the case studies include a review of some of the available material on social and environmental impacts of each dam. Such impacts may include, *inter alia*: impacts on re-settlers, hosts and other affected people, non-economic costs/benefits of migration into or reduced migration out of the region; opportunity cost of lost output from inundated area, salinity and waterlogging; value of lost ecosystem/historical/cultural heritage sites; ecological impacts of flow regulation/reduction; sea water intrusion; losses from reduced numbers of fishes that spawn upstream; losses to some communities due to a change in fishing conditions (while other communities may benefit from these); reduced nutrients in downstream river water; downstream pollution from point and non-point source pollution induced by the dam and negative health impacts.

Comparative Results: Value-Added Multipliers

The comparative values of Value-Added Multipliers for the four case studies, based on a reference set of parameters as defined for each of these studies, are presented in Table 3.3. To understand the impact the changes in values of reference set of parameters, arising as a result of possible changes in underlying assumptions, make on the value of the multiplier, a range of values of the multiplier have been estimated in certain cases. These values are also presented in the Table for purposes of comparison.

The multiplier values, though not strictly comparable across case studies, refer to Type II multipliers that include both inter-industry linkages and income-induced impacts. In the case of the Sobradinho Dam (and the set of cascading reservoirs) in Brazil, the value of the multiplier ranges between 2.28 to 2.40 under the assumptions of the unconstrained supply of labour and capital. However, the multiplier values for Brazil under a supply-constrained scenario for selected sectors, representing the Value-Added Type II multiplier values, are close to 2.00 in most scenarios. This implies that for every unit of value-added generated by the sectors directly affected by the dams, another unit could be generated indirectly in the region.

Table 3.3

Regional Value Added or Income Multipliers in Four Case Studies

Case Study	*Country*	*Methodology*	*Regional Value-Added or Income Multiplier Values under Alternative Assumptions Regarding Project Outputs 'Without Dams'*
Bhakra Dam	Northern India	Social Accounting Matrix (SAM)-based Model	1.78 to 1.90
Bunga Check Dams	Northern India	Social Accounting Matrix (SAM)-based Model	1.41
Aswan High Dam	Eypt	Computable General Equilibrium (CGE) Model	1.40
Sobradinho Dam and the set of cascading reservoirs	Brazil	Semi-Input/Output Model	2.28 to 2.40

Notes: (1) The multiplier values, though not strictly comparable across case studies, refer to Type II multipliers that include both inter-industry linkages and income-induced impacts.

(2) These estimates of multiplier values are under the assumption of unconstrained supply of labour and capital. The multiplier values for Brazil under supply-constrained scenario for selected sectors are around 2.0. For Egypt, the multiplier value is 1.22 under SIM2 with only labour unconstrained.

In the case of the Bhakra dam system the multiplier value of 1.90 implies that for every Rupee (100 paise) of additional value-added in the sectors (agriculture and electricity) directly affected by the project, another Rupee 0.90 (90 paise) are generated in the form of downstream or indirect effects.

In the case of the Aswan High Dam, the value-added multiplier values ranges between 1.22 and 1.40 in the three simulations. In

Simulation III, the multiplier value of 1.40 implies that for every L.E. of value added in the directly impacted sectors, another 0.4 L.E. of value added is generated through inter-industry linkages and consumption-induced effects.

For the check dams in Village Bunga in India, the Value-Added multiplier is estimated as 1.41. Thus, for every Rupee (100 paise) of additional value-added directly by the project in the agricultural sector, another Rupee 0.41 (41 paise) are generated in the form of downstream or indirect effects within the Bunga village and in its neighbouring areas.

Income Distribution and Poverty Reducing Impacts

The comparative results on increases in the consumption/income of various categories/groups of households in three of the four case studies, which permitted such an analysis, *viz.* the Bhakra Dam System, the Bunga check dams and, the Aswan High Dam, are presented in Table 3.4. As already mentioned, the results for Bhakra dam system and Bunga check dams are based on SAM-based, fixed-price, multiplier models while for the Aswan High Dam case study, the results of Simulation 3 have been used for comparison since simulation 3 (unlimited supplies of labour and capital) approximate the SAM-based fixed price multiplier model.

Income Gains for Rural and Urban Groups

In the case of both, the Aswan High Dam and the Bhakra Dam, the per cent increase in income levels, between 'Without' and 'With' project scenarios, have been higher for rural households as compared to urban households though the magnitude of difference in the two cases vary. In the Aswan High Dam, rural households' consumption levels are 22 per cent higher under the 'With Project' scenario in comparison to the hypothetical case where the dam had not been built. The corresponding increase in the case of urban households is 20 per cent. However in the Bhakra Dam case, the per cent difference in gains of the rural *versus* urban households are more pronounced: the total rural households' income increases by 38 per cent under the 'With Project' scenario as compared with a 17 per cent increase in the case of urban households.

Table 3.4

Income Distribution and Poverty Reduction Impacts of Dams

Case Study	*Houshold Category*	*Base (With Project)*	*Without Project*	*Percentage Increase*
Bhakra Dam	Self-Employed Rural Households	12.50	8.83	42
	Rural Agricultural Labour	4.00	2.43	65
	Rural Non-Agriculture	1.13	0.63	79
	Rural-Others	8.41	6.99	20
	Total Rural	26.05	18.87	38
	Urban Households	16.33	14.01	17
Bunga Dam	Marginal Farmers	1206	803	50
	Small Farmers	3015	1891	59
	Medium Farmers	2636	1715	54
	Large Farmers	2516	1762	43
	Workers	1038	842	23
	Total	10411	7013	48
Aswan High Dam	Rural (quintiles):			
	1	8.9	7.4	20
	2	11.6	9.7	20
	3	14.7	12	23
	4	18.3	15	22
	5	31	25	24
	Total Rural	84.5	69	22
	Urban (quintiles)			
	1	9.1	7.9	15
	2	12.9	11	17
	3	17.5	14.8	18
	4	24	20.1	19
	5	48.2	39.6	22
	Total Urban	111.8	93.4	20

***Notes*:** The units for the Aswan High Dam Case study are in LE bn, for the Bhakra Dam case study are in Indian Rupees (Rs). Bn and in the Bunga Case Study are in thousands of Indian Rupees (Rs) Conversion factors are as follows: In Egypt, in 1997, US \$ 1= 3.4 LE; In 2000, US \$ 1= 4.5 L.E. In India: US \$ 1= Rs 10 in 1979-80 and US \$ 1= Rs 45 in 2001-02

Income Gains across Different Categories

While both the rural and urban households in both the Aswan High Dam and Bhakra Dam gain, the gains have not been equal across different categories of rural and urban households. Thus, in the case of the Aswan High Dam, within rural household category, though the percentage difference in gains amongst the different categories is not very large, the third and fifth quintile have been the highest gainers. For urban households the gains to the fifth quintile households have been larger than the others. In the Bhakra case study, rural non-agriculture households gained the highest followed by agriculture labour households. In the Bunga case study, small farmers had the largest increase in consumption as compared to the other four groups of households.

Income Gains for Lowest Income (Poorest) Groups

In the 'With' project situation in the Aswan High Dam case, the income gain for the lowest 20 per cent (first quintile) of the rural population have been 20 per cent. This gain is slightly less than the average rural income increase of 22 per cent. Similarly for the urban lowest 20 per cent of households, the increase in income has been less than the average income increase of the entire urban population. As against an average income increase of 20 per cent for the entire urban population, the increase for the urban poorest group has been 15 per cent. Thus, the gains for the poorest groups in both the rural and urban population have been somewhat less than the average respective gains for the entire rural and urban population groups.

In the Bhakra case study, the income of the poorest group, the rural agricultural labour households, have increased by 65 per cent. In contrast, the corresponding increase for self-employed rural households (the farming household group) has been 42 per cent. Thus, the estimated gains for the poorest group have been much higher than the farming households.

In the case of Bunga, the poorest of the households are represented by three categories—the workers, the marginal and small farmers. The three poorest groups have gained differently. While the workers registered an increase in income of 23 per cent, the income gains for marginal and small farmers have been of the order of 50 per cent and 59 per cent respectively. In contrast, the average increase in incomes of the entire population has been 48 per cent. Thus, while for two of the three groups of poorest households income increases have been higher than the average, for the third group this increase has been somewhat lower than the average.

4 Quantifying Indirect Economic Impacts

Indicative Guidance Notes for Practitioners

R.P.S. Malik and Alessandro Palmeiri

The main objective of this Chapter is to provide in a concise form a more focused and structured exposition to the issues related to the need for proper accounting of the indirect economic benefits of dam projects, methodological issues relating to estimation of multipliers, the nature of data requirement and sources of data availability, and the interpretation and use of the derived results on multipliers. To a large extent the discussion in this Chapter reiterates and/or complements the discussion on some of these issues in the previous chapters and the details on some of these concerns presented in individual case studies detailed in Part II of this book. This Chapter may be viewed more in the nature of an indicative information and assistance guide and scoping of the issues related to the estimation and use of multipliers from the perspective of a water resource development practitioner. This Chapter is however, not intended to be a comprehensive reference on the subject, for which relevant references on different components of the discussion as also the details contained in individual case studies presented in this volume would need to be referred to. The discussion in the Chapter is organised along the following issues:

1. What are indirect economic impacts and why accounting for them is important?
2. How indirect economic impacts can be measured?
3. How to carry out 'indirect economic impact analysis' of projects?
4. How to build a regional SAM?
5. What are the data requirements and sources of data availability?

6. Factors affecting the level of multiplier value
7. How to interpret multiplier values?
8. Analysis of income distribution and poverty reduction impacts of dams
9. Indirect economic impact analysis and project appraisal

What are Indirect Impacts and Why Accounting for them is Important

Investments in infrastructure projects in general, and investments in dam projects in particular, generate a vast array of impacts both in the region where they are located, and at an inter-regional, national and sometimes even at the global level. These include socio-economic, health, institutional, environmental, ecological, and cultural impacts. While some components of these impacts are a 'direct' consequence of the project, others are more in the nature of 'indirect' or multiplier impacts, though there is often some disagreement on what constitutes direct *versus* indirect impacts. Numerous studies have underlined the importance of and difficulties in evaluating a number of these impacts. Of these, one of the main concerns that have often been emphasised is the need to extend consideration to indirect economic benefits and costs of dam projects.

Direct impacts of a dam project are measured in terms of value of additional output of agricultural commodities, hydropower, navigation, drinking water, tourism and reduction in flood damages. The changes brought about by direct impacts generally alters the nature and pace of the prevailing state of affairs in the project and its surrounding region and through a 'ripple' effect provides impetus to the economy resulting in increased overall economic activity in the region. Indirect and induced economic impacts are those that stem from the linkages between the economic and other direct consequences of a dam project with the rest of the economy. Thus, the indirect economic impacts include: (i) inter-industry linkages, backward and forward, resulting in an increase in the demand for outputs of other sectors, and (ii) consumption-induced impacts arising from additional incomes generated by the dam project. Accounting for such indirect impacts of a project is important in facilitating more informed decisions relating to funding of the project as also its subsequent evaluation, more so for projects constructed in the public sector, though these considerations may not carry much weight for an entrepreneur wishing to set up such a project.

While direct impacts are obvious and usually easy to identify and not too difficult to quantify, indirect and induced effects need some elaboration. To illustrate, water released from a multipurpose dam provides irrigation for increased output of agricultural commodities. Changes in the output of these commodities require inputs from other sectors such as energy, seeds, fertilisers, etc. Further, increased output of some agricultural commodities encourages setting up of food processing and other industrial units. Similarly, the hydropower produced from the multipurpose dam provides electricity for households in urban and rural areas and for increased output of industrial products (e.g., fertilisers, chemicals, machinery). Changes in the output of these industrial commodities require inputs from other sectors such as steel, energy, and chemicals, among others. Thus, increased output of both irrigation and electricity from a dam results in significant backward linkages (i.e., demand for higher input supplies) and forward linkages (i.e., providing inputs for further processing).

Increased outputs of industrial and agricultural commodities generate additional wages and incomes for households. Higher incomes result in a higher consumption of goods and services that, in turn, encourage production of various agricultural and industrial commodities. Further, changes in output generated by the project may affect prices of direct project outputs, inputs, substitutes, complements and factors of production. Changes in wages and prices have both income and substitution effects on expenditure and saving decisions of different owners of factors, which further impacts the demand for outputs both within the region and throughout the economy. Induced impacts reflect the feedbacks associated with these income and expenditure effects, and include any impacts of changes in government revenues and expenditures that resulted from the project. Such indirect effects of a project though easily recognised and often appreciated and discussed are generally not accounted for due often to hazards associated with their quantification.

While the direct impacts generally start flowing immediately after the project is commissioned and the water and electricity starts flowing from the multipurpose dam, most of the indirect[1] impacts follow the direct impacts with a time lag. The extent of this time lag however, varies depending upon initial underlying conditions prevailing both in the project region as also at a more aggregate level in the economy.

How Indirect Economic Impacts can be Measured

Indirect effects can be measured through estimation of multipliers. Multipliers are summary measures expressed as a ratio of the total effects (direct and indirect) of a dam project to its direct effects. A multiplier of 1.75, for instance, shows that for every one dollar of value-added generated directly by a project at maturity; another 75 cents are generated in the form of indirect effects. Two types of multipliers are widely utilised. The so-called Type I Multipliers account solely for production-related impacts due to inter-industry linkages, while Type II Multipliers account for both production-related and consumption-induced indirect effects. Here we will however, be concerned with Type II multipliers only.

Estimation of the multiplier thus, requires a careful analysis of the direct effects. This essentially involves quantification and valuation of major outputs of the dam and the assessment of the share of direct effects that is attributable to the dam project. As shown in Box 4.1, for estimating a project multiplier value for a dam, for the numerator, we need to estimate the regional value-added under 'With Project' situation

Box 4.1

Values of Variables Required for the Estimation of a Project Multiplier

Definition of Project Multiplier	Regional Value-Added with Project **minus** Regional Value-Added without Project
	Value-Added of Agriculture and Electricity with Project **minus** Value-Added of Agriculture and Electricity without Project

1. There could be some indirect economic impacts during the construction phase of the project such as investments in manufacturing of cement. For the time being these are not being considered.

as well as the regional value-added under 'Without Project' situation. For the denominator, we need to estimate the value added from the sectors that are directly affected by the major outputs of the dam (namely agricultural output, hydro electricity, water supply etc.).

Apart from identifying the indirect economic impacts and defining a measure for their quantification, there are a number of other important related questions that often merit consideration in both design of a project and its evaluation. The most significant of these considerations relate to the distribution impacts (direct plus indirect) of a dam project and are largely localised, that is, concentrated mostly in the geographical region surrounding the project site, or these are spread out to surrounding regions and even beyond. Who reaps the benefits of these impacts? Are these limited to the rural population or does the urban population also gain? Within the rural populations are these benefits limited to large landowning households alone or other classes of households, including small and marginal farmers and landless labourers, also gain? What is the relative importance of direct and indirect impacts in increased incomes of different classes of households? Analysing the indirect impacts of a dam project thus enables some additional insights to the linkages and pathways through which the benefit of the project flow to different sections of the society.

Dam projects in addition generate a vast array of other impacts—social, health, institutional, environmental, ecological, and cultural. Such impacts may include, *inter alia* social, economic, and cultural impacts on re-settlers, hosts, and other affected people, including indigenous people; non-economic costs/benefits of migration into, or reduced migration out of, the region; opportunity cost of lost output from inundated area; impacts of salinity and waterlogging; value of lost ecosystem/historical/cultural heritage sites; the ecological impacts of flow regulation or reduction; sea water intrusion; losses from reduced numbers of fishes that spawn upstream and due to changes in fishing conditions; reduced nutrients in downstream river water; downstream pollution from point and non-point source pollution induced by the dam, and positive or negative health impacts. While some of these impacts are direct consequences of the project, others are more in the nature of indirect impacts. Further while some of these impacts are intended, there are a number of unintended impacts as well. Accounting for such a large number of multi-sectoral direct and indirect, intended and unintended, negative and positive, inter-temporal and spatial impacts of dams poses a challenge for their effective assessment. While not undermining the importance and desirability of accounting for all these impacts in a comprehensive assessment of the impacts of dam projects, it has however, not been possible to account for all of such impacts due to methodological challenges related to quantification of most of these impacts. The work reported in this book on estimation of the "multiplier", a measure that reflects the production-related and consumption-induced indirect economic impacts of dams, is an attempt to move just one step forward in our efforts towards achieving the broader goal of making a more comprehensive assessment of dam projects. Quantification of some of the other impacts and their consideration in *ex-post* and *ex-ante* analytical frameworks is highly desirable and would go a long way in further improving our understanding of the impacts of dams. Efforts need to be initiated towards this end.

How to Carry Out 'Indirect Economic Impacts Analysis' of Projects: Analytical Tools

The analytical tools that can be employed to analyse indirect and induced impacts of exogenous changes fall into the broad category of multi-sector models. A range of such models exist, differing primarily with respect to (i) their assumptions regarding quantity *versus* price-responsiveness to exogenous shocks, (ii) the focus on income levels *versus* the inclusion of distributive considerations, (iii) their capacity to incorporate factor and import substitution possibilities, and (iv) their capacity to accommodate policy distortions, specific factor and output market structures, and other peculiarities. The design and application of modeling techniques that compute multipliers can be useful in both understanding and underscoring the pathways of benefit generation associated with a project, as also to understand the interactions between a project's output and the structure of the economy and institutional/ policy setting in which it is introduced.

The following economy-wide, multi-sector models have been suggested in the literature for estimation of multiplier effects and to perform distributional analysis of dam projects:

i. Input-Output (I-O) and Semi-Input Output (S/ I-O) Models;

ii. Social Accounting Matrices (SAM) based models, including the IMPLAN Models; and

iii. Computable General Equilibrium (CGE) Models

On the face of it, it might seem as if the analyst has a choice of the above three completely different types of models, each with its individual advantages and disadvantages. However, it may be important to mention that these models are not strictly independent, but are to a large extent an extension or variant of each other. The models, in some instances, are also linked in that the output of one model forms the input of the other. To ensure that the analysis carried out estimates the full economic impact of the project, wherever possible, these models should be used in a complementary fashion.

A crucial distinction between SAM-based multiplier analysis and I-O models is that SAMs are able to account for the way in which initial asset distribution and factor endowments—and therefore the distribution of income among household categories and between them and the corporate sector—interact with the structure of production in determining final outcomes, particularly for welfare analysis.

The SAM embodies the fact that within the macro-economy there is a circular flow process and that what happens at one point on the circuit will have implications for experience at other junctures, so that one needs to be equally concerned with all the different aspects of technology and behaviour that together describe the circular income flow and the connections in the economy. SAMs also have the potential to provide a powerful and flexible framework for a systematic and integrated approach to national accounts at the appropriate level and type of disaggregation.

While each of these models can be used for estimation of indirect and induced effects, however, SAM-based or CGE modeis are better suited for the analysis of income distribution impacts of a dam project.

When prices are assumed fixed, as in I-O models or SAM-based multiplier analysis, all adjustments occur through quantity changes. A change in demand for a sector's output results in changes in that sector's output supply and factor demand. In the absence of supply constraints, adjustments occur via impacts on labour or capital employment and inter-regional factor migration. The presence of idle labour or capacity somewhere in the system—either locally or in other regions, if the model is inter-regional—prior to the exogenous shock is thus crucial for the existence of quantity-driven multiplier impacts as estimated by these models.

On the other hand, a variable-price model, such as a standard CGE, implies the presence of supply constraints, so that for at least one of the factor the aggregate levels of factor employment is fixed. In this case, a change in sectoral demand results in relative price changes, thus determining substitution effects among inputs and among outputs, with factor relocation across sectors in the regional economy. CGE models, however, can be flexible enough to incorporate alternative specifications of factor mobility and wage differentiation, and have successfully been modified to solve as I-O models, so that comparisons can be made between the welfare impacts under alternative assumptions in this area. The greater flexibility of CGE models therefore suggests their use whenever possible based on data and human resource availability. SAM based multiplier analysis, characterised by the assumption that the prices of commodities, factors and foreign exchange are all fixed, is a special case of CGE analysis.

From the point of view of the analysis of indirect and induced impacts of dam projects, however, the choice of analytical tool should not always favour the most sophisticated one, but rather be driven by the assumptions regarding the mechanisms through which impacts are transmitted in the specific region of interest, and particularly regarding factor mobility. The utility and choice of type of model for estimating the multiplier would depend, apart from the objective of analysis, on several other factors, including on data availability, availability of technical expertise, its utility for future applications and the knowledge base that it generates. For projects with regional development objectives, SAM based models appear to be more appropriate and doable. Similarly for social impact assessment of water projects, especially in developing regions, a SAM based model appears to be quite a promising tool. However, since SAM and I-O tables are built on data for a given point in time (e.g. for the year 1990), they provide only a snapshot view of the economy for the year for which SAM is built. Therefore, these models are not capable of providing forecasts for economic development over a period of time. In order to obtain an idea of future developments a computable general equilibrium model must be employed. Although CGE models do also rely mainly on one point in time data they do contain functions, such as production and consumption functions that bridge the time gap. In reality these functions form only a substitute for the missing statistical time-series.

It is important to note some of the limitations of each of these methodologies for estimation of multipliers so that appropriate caution is taken when using these methodologies and interpreting results based on them. As discussed above, the main limitation of the I-O model is that it assumes linearity in production and cost-determined prices independent of demand. I-O models are basically planning-oriented models, based on the assumption that quantities on the supply side of the economy will respond to exogenous shocks on the demand side in the absence of any supply-side constraint. Supply functions are perfectly elastic (horizontal, instead of the more realistic upward-sloping shape reflecting the scarcity of one or more inputs), and both output and input prices are unaffected, regardless of the size of the exogenous shock. Most of the limitations characterising I/O models also affect SAM-based multiplier analyses. However, SAM based models represent a step forward in that they close the loop between factor incomes, their distribution to different endogenous and exogenous economic agents, and their expenditure behaviour. Since I-O and SAM models completely ignore supply side constraints, these unconstrained models may exaggerate the size of inter-sectoral linkages.

CGE models to a large extent take care of some of the concerns of I-O models such as those relating to price endogenity, price responsive input and output substitutability, factor supply constraints which generate output supply constraints etc. However, most empirical applications of CGE models have been developed on the simplifying assumption of constant returns to scale production technology and perfectly competitive market structures. Like the I-O model, the comparative static CGE model does not contain any explicit time dimension, though a recursive-dynamic CGE model can be linked to a macro-econometric model to produce a 'business-as-usual' forecast. In general, CGE models require considerable data, which, in most cases, is difficult to obtain. This problem is more severe at the regional level, where data in most cases is virtually non-existent. In fact, one of the possible reasons for the relatively slow start of regional CGE modeling is the paucity of regional data, in addition to unresolved theoretical issues of regional specification.

Another concern with the use of these models relates to their validation and calibration. Calibration or benchmarking determines the values of the normalising parameters so as to replicate the observed flow values incorporated in the model. Thus once the model has been specified, the parameters in these algebraic equations must be evaluated. As mentioned earlier, data on endogenous and exogenous variables obtained at a given point in time are typically used for this purpose. Since the evaluation of model parameters is done on the basis of a single observation, there is often a concern about the reliability of the results obtained from the model. Substantial efforts have to go in to make sure that once all the parameters are specified, the model when solved should reproduce the benchmark scenario.

How to Build a Regional SAM

SAMs can be constructed in a variety of ways. The manner in which a SAM is specified is typically driven by the problem being addressed. Although there is no specific sequence in which to proceed with the construction of a SAM, however, since SAM is an extension of input-output matrix, the important starting point for the construction of a SAM is the accessibility of an input-output matrix. Quite often the input-output matrices at the national level are available, however these are generally prepared at a gap of almost five or sometimes more years. For example, in a country like India, which has an excellent statistical database and enough trained manpower, the latest I-O matrix available at the national level relates to the year 1998-99. Supporting policy analysis through development and analysis of economy wide models would ideally require access to the latest economic data. However updating an old I-O table to a recent year is both a difficult and a challenging job more so in developing countries due to insufficient and fragmented data availability, that too with a lag, as well as concerns arising out of reliability of the available data.

In most of the developing country situations it is however, not common to find I-O matrices at a sub national level—at the level of a region or a State. Wherever available at the State level these have generally been constructed as part of some research study but not by any official agency. For example, to the best of our information, in India the latest I-O matrix for the state of Tamil Nadu is available for the year 1981-82 while for the state of Punjab the only year for which an I-O table is available is for the year 1979-80, that too was constructed in 1990 as part of an IFPRI study (Bhalla *et al.,* 1990). Building an I-O matrix, more so at the regional level, is a very data and manpower intensive operation and, therefore, where available, it is better to start building a SAM from an existing I-O table. In case an I-O table for the region of interest is not available, it may be desirable to either compile an I-O table, if the required data for the region is available, or else use, as a first approximation, the I-O table from a region, whose structure of inter-industry linkages is assumed to be sufficiently close to that of the region in question.[2] Before doing so, it is however, important to define the region over which most of the benefits of the project extend, though in practice it may sometimes be difficult to correctly define the boundaries of this region.

Defining the Region's Boundary

The region over which the impacts of a dam project may extend depends primarily on the size (small *versus* medium *versus* large) and nature (single purpose *versus* multipurpose) of the project. While the major direct impacts of medium and large projects are generally felt at the sub–regional or regional level, the major benefits of small projects are often

2. There do exist cases where SAMs have been built in the absence of I/O tables, a recent example exists for Mozambique (Arndt *et al.*, 1998).

more localised and may sometimes be restricted to the level of a village. However, the indirect impacts of the dam project may extend to a much larger area—sometimes to the national level and beyond. For example, in the case of the Bhakra multipurpose dam located in the Punjab state in India, while the direct benefits in terms of irrigation water, drinking water and power benefits are shared by four other neighbouring states—Haryana, Rajasthan, Delhi and Himachal Pradesh, the indirect impacts of this dam extend to the entire country. High demand for agricultural labour in the Bhakra command area attracts migrant labour from far off states of Bihar and eastern Uttar Pradesh. The income remittances by these migrant labourers to their native villages in far-flung poor regions has helped improve the living conditions of people living hundreds of miles away and has created its own further downstream effects. A majority of the migrant labourers utilised these savings for purchasing consumer durables such as television, radio, bicycle, sewing machine etc. Apart from earning cash, these migrant labourers have also acquired new skills in crop cultivation and operation of farm machinery during their stay in Punjab, which has helped create its own downstream effects in their native villages. Further, the increased foodgrain production in the Bhakra dam command has helped in enhancing food security for the entire country, resulted in lower imports of foodgrains and lower food prices especially for the urban poor. Thus, the impact of the surplus foodgrains from the Bhakra dam has been felt by people in areas far away from the location of the dam. Similarly in the case of the Sobradinho dam in Brazil, the area of influence of the cascade of large reservoirs along the Sub-Médio São Francisco goes well beyond the immediately surrounding region and development from the dams affects output, incomes and consumption levels for the whole Northeastern region. In contrast, the direct economic impacts of small check dams in village Bunga have been limited mainly to the village itself while some indirect impacts extend to the urban areas in its immediate vicinity.

Thus, in estimating the multiplier effects of dams, especially where a regional or sub-regional perspective is considered important, it is essential to carefully define the region that surrounds the project or is affected by the project. The model structure and related databases should present the appropriate regional characterisation so as to incorporate as much of indirect effects as possible. Quite often where the impacts are felt in far off regions or in regions which are not in the geographical contiguity of the region affected directly by the project, these studies need to be supplemented by other evaluations that may have either been carried out as part of some other studies (e.g. the impact of the Bhakra dam on food prices for the urban poor and the impact of remittances in villages in eastern Uttar Pradesh and Bihar) or would need to be undertaken specially as part of such studies.

Defining the Level of Disaggregation

In addition to defining the boundaries of influence over which the direct and indirect impacts of a dam project may extend, the SAM structure should incorporate the degree of disaggregation that is desired to be incorporated in consonance with the objectives of building the SAM and the availability of required data. The SAM structure is flexible enough to incorporate the desired level of disaggregation. Thus, for example, if the SAM is to support analyses of dam projects, it must include a detailed disaggregation of agricultural factors, activities and commodities as well as the other areas, such as electricity production, that are most directly affected by the dam. Similarly, in order to carry out a meaningful welfare and poverty impact analysis of the consequence of a dam project, it is important that the SAM structure is sufficiently and meaningfully disaggregated in its representation of households, so that the groups that stand to lose or gain from the project can be identified.

What are the Data Requirements and Sources of Data

A SAM is a square matrix that, for a period of time (typically one year), accounts for the economy-wide circular flow of incomes and payments. It summarises the structure of an economy, its internal and external links, and the roles of different actors and sectors. A SAM brings disparate data into a unified framework.

The specific data requirements for constructing a regional SAM vary depending on the type of problems being addressed. While it is not possible to strictly identify a unique set of required data for building a

SAM, some generalisations however, can be made. In addition to standard input-output data (industry production, inter-industry transactions, final demands, factors of production and imports/exports), typical SAMs require additional data on total factor payments, total household income (by income category), total government expenditures and receipts (including intergovernmental transactions), institutional income distribution, and transfer payments (both to households and to production sectors). The databases and sources of data required to build a SAM and estimate multiplier values also depends upon the degree of disaggregation that is incorporated in the SAM. More specifically the data required for building a SAM can be divided into six broad groups:

1. National accounts.
2. Balance of payments.
3. Monetary accounts.
4. Public sector budget.
5. Input-output matrix.
6. Data on household consumption, factor employment and capital stock.

Diverse publications of the Central and State governments provide a rich source of secondary data. This often needs to be supplemented by primary data on several aspects including consumption expenditure surveys carried out either as part of the official data collection efforts or specially carried out. Data on certain aspects, such as inter-regional trade, especially by road, are often not routinely collected and may need to be collected especially if the model is inter-regional. Since SAMs are typically built as static snapshots of a region, the data elements will need to be generally consistent in temporal and geographic specificity

IFPRI's Trade and Macroeconomics Division's web page (*www.ifpri.org*: click on Research, then Research Divisions, then Trade and Macroeconomics, then Divisional Output, then Discussion Papers) presents a number of studies, downloadable in Acrobat format, that describe data sources and methods that were followed to build SAMs in a number of countries, often with very sparse data.

Data Requirements for Building a Village Level SAM

Like a national or regional SAM, a village SAM has also to capture the linkages amongst production activities, institutions and outside world. While the nature of data requirements for building a SAM, be it at the national level, a regional level or even at a village level are almost identical, the sources of data could vary substantially. However, unlike at the national or regional level where available official data supplemented by ad-hoc survey data can be used to construct the SAM, no such ready made data sets are available in the case of a village. The only way out is to collect all the required data through appropriately designed household sample surveys supplemented by participatory rural appraisals (PRA), survey of local institutions, and some secondary data that may be available with local level government functionaries. Since building a SAM requires, apart from survey data, a complete understanding and knowledge of the dynamics of various transactions taking place in a local economy this may require personal indulgence of the SAM builder and is therefore, generally a time consuming process. In the case of estimation of multipliers for Bunga check dams, reported in Chapter VII of this study, we adopted such an approach to collect the requisite data to build the village SAM and trace the indirect impacts of check dams constructed in the village.

How to Balance a SAM

Since the construction of the SAM requires a consistent consolidation of data on various aspects collected from different sources and often differing in their year of reference, it is not surprising to find imbalances between the row and column sums for most accounts. These accounts need to be integrated and balanced to build the SAM. IFPRI has developed a cross entropy estimation approach that is very useful in balancing SAM. The General Algebraic Modeling System (GAMS) programme found as part of the standard CGE model at *www.ifpri.com* checks that the SAM that is entered is balanced, that is, for each account, the row and column totals are equal. If the absolute value of the sum of account imbalances exceeds a cutoff point, an optimisation programme is used to estimate a balanced SAM. The programme, which minimises the entropy distance of the cells of the estimated SAM from those of the initial SAM subject to the constraint that row and column totals are equal, is primarily intended at removing rounding errors (see Robinson *et al.*, 2000)

How Much Does it Cost to Carry Out Multiplier Analysis

Constructing the SAM and performing the macroeconomic analysis using SAM based or CGE models is an expertise and data intensive operation and therefore requires substantial upfront investment to undertake such an analysis. The time and cost of undertaking such an exercise depends on several factors—the size and nature of the project and the radius of its influence both in terms of its direct and indirect impacts, the data availability, the extent of new data required to be collected, the availability of an I-O matrix for the region of concern, the availability of computer hardware and software for the purpose, the availability of trained manpower, the objectives of analysis and the level of disaggregation and sophistication required to be incorporated. Since this type of analysis is often not carried out either at the project appraisal stage or for ex-post evaluation of projects, or if carried out is not separately costed, it is difficult to estimate *a priori* the cost of undertaking such an exercise—either in absolute amount or as a fixed percentage of project cost. Some order of magnitude numbers are however, available from a South African project (Mullins, n.d) even though these numbers may not convey much and may be off the mark by any percentage amount. Building the SAM for the Komati Basin project in South Africa cost US$ 120,000 (on a US$ 470 million project) and required specialised knowledge in national accounts and computer software. In terms of percentage it cost just 0.03 per cent of the capital cost of the project. In absolute terms the US$ 120,000 is not overly expensive given that it provides an important tool for assessing the project relative to its objectives.

The work reported in this book on estimation of multipliers for four case studies was built upon some already existing data and previous limited analysis available of these cases. On an average each study costed about US$ 75,000. Thus, depending upon the nature and quality of data available and analysis available from some previous studies that may have been carried out, carrying out an *ex-post* indirect economic impact analysis of a multipurpose dam project could cost anything between US$ 50,000 to $ 100,000.

However, if the cost and data considerations so demand, the construction and use of models such as SAMs may be undertaken in a step-wise fashion in terms of the reliability of their results. While the model should ideally be fully defined at inception, further disaggregation, refinement and reliability of results may be achieved in stages through incremental investment in the data collection and analysis that underpins the model.

Factors Affecting the Level of Multiplier Value

An understanding of the factors affecting the level of multiplier value is important in designing the project and taking corrective actions wherever required to help increase the effectiveness of the project in achieving desired objectives. The value of the multiplier is likely to be affected by several factors—some of the important factors that may affect the magnitude of the multiplier value include:

- Size and nature of the project—generally the level of multiplier values are expected to be higher for a large dam *vis-à-vis* a small dam and for a multipurpose *vis-à-vis* a single purpose dam due to the varied nature and quantum of backward and forward linkages and consumption-induced indirect effects[3] associated with different types of projects.
- Level of diversification in the regional economy—the more diversified the economy is, the higher is likely to be the value of the multiplier. It is also expected that the level of multiplier values will be higher for a large dam providing water for multiple-crops compared with those for a single-crop system (e.g. the Muda project with paddy monoculture). For example, if outputs

3. Care should be taken, however, in identifying the trade-offs between potentially competing uses—e.g., irrigation *versus* hydropower, when timing of release are not consistent—that may dampen project benefits.

include oilseeds, sugarcane and cotton, there will be investments in agro-processing units, sugar mills and cotton ginning and processing mills and/or textile units.

- Level of inter-sectoral integration—the higher the level of integration, it is likely that the higher is the value of multiplier
- Level of inter-regional integration—multiplier value is likely to be influenced by the level of dependence from inter-regional demand
- Policy environment and institutional regime at various levels—multiplier value is likely to be affected by policies and institutional regimes influencing incentives, such as those relating to input use and crop selection, land tenure systems, subsidies for inputs such as water, electricity, fertilisers and new agricultural machinery, availability of credit, free movement of labour and goods across regions, provision of support prices for crop outputs, incentives for setting up of industries etc. These policies when complemented by an efficient institutional regime such as that relating to ensuring timely provision of required inputs for agriculture, efficient markets for crop disposal, provision of infrastructure such as rural roads and transport, mechanism for efficient allocation of water and electricity produced by the dam amongst competing uses etc are likely to influence the magnitude of the multiplier.
- The level of household incomes and the pattern of consumption, and propensity of consumption of households.
- Phase of implementation of the project—the level of indirect impacts and their nature will also differ during different phases of the implementation of a dam project. During construction, the project will require construction materials generating backward linkages with various supplier industries. During this phase, large numbers of workers will be engaged, some of whom would remit funds to other regions of the country. The multiplier levels will also be relatively high during the operational phase of the dam when full benefits of the project are available since backward and forward inter-industry linkages, as well as consumption-induced effects will be quite significant.

How to Interpret Multiplier Value

Multiplier value provides information on the level of indirect impacts in relation to the direct impacts of a project. As already explained a multiplier value of 1.75, for example, implies that for every one dollar of direct benefits contributed by the project, another 75 cents are generated in the form of indirect impacts. The analysis however, should not focus solely on the numerical values of multipliers. One has to look at how the indirect effects are transmitted in the economy—what matters is an understanding of the nature and generation process of regional economic impacts and through what pathways are these transmitted. This process may help identify not only the channels through which the majority of benefits and costs are produced (direct *versus* indirect *versus* income-induced), but also potential additional outputs or uses thereof—as well as unforeseen trade-offs and negative impacts.

Similarly multipliers should not be read as indices of a project's desirability. It should be kept in mind that a lower value of multiplier does not necessarily mean that the project is inferior in comparison to another project whose value of multiplier is higher. Indeed, a small multiplier may be associated with a highly beneficial project that generates primarily a direct impact. On the other hand, large multipliers may translate into large overall net benefits depending on the initial importance of the directly impacted sectors in regional GDP and employment. The relative importance of different factors in influencing the level of multiplier values needs to be assessed by undertaking a sufficient number of case studies.

Notwithstanding the exact magnitude of the value of multiplier, an important message that the presence of multipliers conveys is that, assessing the project based solely on its direct impacts may be missing a sizeable portion of its benefits.

Analysis of Income Distribution and Poverty Reduction Impacts of Dams

SAM-based models and CGE models explicitly take income distribution considerations into account through a disaggregation of households into various income categories. Such a disaggregation enables carrying out a meaningful welfare and poverty impact analysis of the consequence of a dam project and to clearly identify groups that stand to lose or gain from the project. This also enables tracing the direct and indirect channels through which such gains and losses take place.

In the case of Bhakra dam, for example, the results on income distribution show that the gains to the agricultural labour households from the dam were higher than gains to other rural households and to urban households. For the agricultural labour households, the income level under 'with project' situations are estimated to be 65 per cent higher than the income level under the 'without project' situation. The corresponding figures for self-employed rural households and urban households were 42 per cent and 16 per cent, respectively.

The SAM-based, fixed price, multiplier model for the Bhakra dam has also been used to estimate direct and indirect income effects of additional irrigation and hydropower provided by the Bhakra dam for each of the household categories. The resulting estimates show, that a major part of the income of rural households comes from the output of sectors directly affected by the dam. For example, for self-employed rural households, about two-thirds of the difference in income under 'Without Project' and 'With Project' situations are due to higher output and value-added in sectors that are directly affected by the dam, namely agricultural crops and electric power. The rest, about one-third, of the difference in income is attributable to changes in the value of output and value-added of sectors that are affected indirectly (through inter-industry linkages and consumption-induced effects). In the case of agricultural labour households, as much as 70 per cent of the difference in income is due to changes in output/value-added of sectors that are directly affected by the dam. However, in the case of urban households, the situation is completely reversed. As much as 83 per cent of the difference in income (under with and without project situations) is accounted for by changes in the outputs of sectors that are indirectly affected by the dam, namely agro-processing, manufacturing of textiles etc.

The findings are of importance to project analyst intent on improving the distribution as well as the absolute levels of incomes in the project area, for it again implies that the indirect benefits cannot be easily ignored. This also underlines that much more attention than is typically given to project design and appraisal work needs to be given to complementary policies that might help ensure a more equitable distribution of these secondary benefits.

Indirect Economic Impact Analysis and Project Appraisal

A question that needs to be resolved is which of the economic impacts should be included in the project appraisal and which in *ex-post* evaluation, and for what purposes. *Ex-ante* dam project appraisal generally tends to focus solely on the value-added produced by the sectors directly affected by the project, with no consideration for the 'multiplier effects' their outcomes produce on the regional, national, and sometimes global economy.

As per the standard approach of project analysis, currently used by practitioners, multipliers are generally not used for project evaluation because of the difficulty of avoiding double counting of benefits. It is argued that shadow prices that include carefully traced indirect changes in value-added include the multiplier effects while minimising the danger of double counting. Behind this argument are the two key underlying assumptions in using shadow prices that justify such an exclusion of multiplier effects from consideration. These conditions stipulate that the full employment conditions hold and the project under consideration does not alter relative prices.

In practice however, the two standard assumptions may not hold, as may well be the case in a number of agricultural and industrial sectors in developing countries where dams are introduced. As a result, social cost-benefit analysis may not fully capture a project's net benefits because multiplier impacts on output, income and employment can indeed be large and should not be ignored. In addition the economy-wide models also prove useful especially when distortions that may affect project outcomes are associated with policies whose impacts are not traditionally quantified in cost-benefit analyses (CBA), such as macroeconomic policies. For large infrastructure projects whose impacts may affect the productive structure of a region and, conversely, can be affected by exogenous shocks and policy changes, this type of analysis can be a useful complement to traditional CBA. In addition, these techniques can be applied to estimate region-specific shadow prices, which would be needed for appropriate CBA but seldom exist.[4]

Even if the two basic assumptions were to hold good, for large-scale projects, regional and macroeconomic impacts may have potentially large consequences in terms of income distribution, spatial configuration of regional development, and the non-economic impacts on regions other than the project area. These are often of great interest to project developers or to the government of the region or country where the project is located. While not needed as part of a project's social cost-benefit analysis, these impacts need to be adequately evaluated and presented to decision-makers, so that they can make a more informed decision based on the broader set of consequences produced by a project and its alternatives.

Having underlined the unavoidability of using multiplier analysis to complement Social Cost-Benefit Analysis, a question that needs to be resolved is how to use the information on multipliers derived from *ex-post* evaluation of a project as a predictive tool in *ex-ante* appraisal of dam projects. Rather than applying an ad-hoc factor, to take account of multiplier effects, to inflate the value-added produced by the sectors directly affected by the project, it is considered desirable that valuable lessons can be learned by assessing the full range of developmental impacts, including the distributional impacts, of dam projects in the context of *ex-post* evaluations, and that *ex-post* evaluation in turn could serve as a valuable tool for the identification and design of alternatives for new projects, as well as for their *ex-ante* assessment.

Indirect Economic Impact Analysis and Project Monitoring

Given the significance of accounting for indirect economic impacts, it would be desirable to not only use this information in project appraisal and evaluation but also to monitor the project during its implementation phase so that necessary corrective actions could be

4. See Bell and Devarajan (1980).

taken to enhance the growth linkages and in spreading the benefits to wider sections of society.

Measuring the progress of the project, and its subsequent evaluation, has implications for data collection. For an effective monitoring and evaluation, data collection on base line conditions should commence even before the project starts. Subsequent data collection after the construction of the project work starts should continue at regular intervals not only until the project is completed but also subsequent to that till most of the adjustments on account of the project are complete. The data collection efforts must approach with an analytical framework that attempts to trace direct as well as the indirect impacts of the project not only in the periphery of the project site but beyond in areas in which the indirect impacts may extend.

Since the indirect impacts can be many and varied, monitoring the indirect impacts requires collecting relatively huge amounts of data on a wide variety of variables at different levels of aggregation and for a relatively much longer period of time. As mentioned above, most of the indirect impacts emanate essentially as a result of changes in variables directly impacted by the dam. For monitoring the indirect impacts and more clearly understanding the linkages between the direct and indirect impacts, one would therefore, need to collect the baseline data both on the variables that are likely to be impacted directly from the construction of the dam as also of the variables that are *a priori* expected to be impacted indirectly. Since the Government policies and a host of other complementary variables (such as the availability of railroad infrastructure, availability of credit, technology etc.) can play a major role in determining the magnitude of the indirect impacts, in speeding up or slowing down the transmission of these indirect impacts in the economy, and in the distribution of the gains emanating from these indirect impacts across different sections of the society, baseline information on some such variables would also need to be collected. While some of such indirect impacts can occur in the short run, others can be realised only in the medium and somewhat long run. To capture, therefore, some of these complexities, the data collection efforts would need to be extended to a much wider area and for a much longer time period.

Without such a detailed time series of data it is difficult to obtain a real understanding and a quantification of the magnitude of structural changes that have taken place. This multifarious data collection over a long period of time is of course an expensive affair but in the larger interest of improving the efficiency of the project during its implementation and its subsequent evaluation is an indispensable necessity for which necessary funding must be provided at the project formulation stage. Due to non-availability of such detailed data, for estimating project benefits, one has often to resort to the use of models to picture the pre and post project situations in the regional economy. Apart from aiding proper monitoring, the availability of such data could avoid the need for resorting to such modeling to estimate project benefits and in improving the reliability of the results obtained on project evaluation.

References

Arndt, C.A. Cruz, H.T. Jensen, S. Robinson, and F. Tarp (1998). "Social Accounting Matrix for Mozambique 1994 and 1995", *Discussion Paper* No. 28. Washington D.C.: IFPRI

Bell, C. and S. Devarajan (1980). "Semi-Input-Output and Shadow Prices: A Critical Note", *Oxford Bulletin of Economic|s and Statistics*. August.

Bhalla, G.S., G.K. Chadha, S.P. Kashyap and R.K. Sharma (1990). *Agricultural Growth and Structural Changes in the Punjab Economy: An Input-Output Analysis*. Washington D.C.: IFPRI

Gittinger, P. (1982). "Economic Analysis of Agricultural Projects", 2nd Edition, *EDI Series in Economic Development.* Baltimore, MD: The Johns Hopkins University Press.

Mullins, David (N.d.). *Macro Economic Analysis*. Contributing Paper Prepared for Thematic Review III.1. World Commission on Dams. *www.dams.org*

Robinson, S., A. Cattaneo, Moataz El-Said (2000). "Updating and Estimating a Social Accounting Matrix Using Cross Entropy Method", *TMD Discussion Paper* No. 58. Washington D.C.: IFPRI

5 Recommendations for Additional Work

Ramesh Bhatia, R.P.S. Malik, Monica Scatasta and Rita Cestti

As we have seen from the results of the four case studies, indirect economic impacts are quite significant in relation to the direct economic effects of dams. Further, these projects also provide substantial benefits to people who are in the relatively lower income groups. Hence, it is critical that such indirect economic impacts are explicitly considered and evaluated for water infrastructure projects.

Further, since water investments are not 'marginal' or small in comparison with the rest of the economy, it is important to capture the total benefits of the project in the context of regional development. Such considerations of creating employment and regional incomes were the motivation behind large-scale water and other infrastructure projects in the United States, Europe and other countries. Regional benefits were the reasons that major projects such as the Hoover dam, Grand Coulee dam and the Columbia Basin Project were undertaken with federal funds in the United States. As pointed out by Briscoe (2003), regional multipliers were precisely the point of such projects and it was expected that water investments would "change the trajectory of regional economies." Multipliers were the explicit and overt reason for these projects.

However, in practice, the Office of Management and Budget in the United States explicitly states that secondary benefits should not be included in project appraisal since secondary benefits resulted in no net economic gain at the national level.[1] *Ex-ante* dam project appraisal tends

1. In the words of one summary (Ortolano, personal communication): "The issue (of indirect benefits) was intensely debated in the early 1950s and, as far as US government agencies are concerned, it was resolved with the distribution of Circular A-47 of the Bureau of the Budget. This document forbade the inclusion of secondary benefits, basically concluding that secondary benefits resulted in no net economic gain at the national level."

to focus solely on the value added produced by the sectors directly affected by the project, with no consideration for the 'multiplier effects' their outcomes produce on the regional, national, and sometimes global economy. It has been suggested by Gittinger (1982), that the multiplier analysis is generally not used in project analysis because of the difficulty of avoiding double counting of benefits. Shadow prices that include carefully traced indirect changes in value-added include the multiplier effects while minimising the danger of double counting.

In view of this difference between the current practice of *ex-ante* project appraisal that ignores indirect economic impacts and the significance of the indirect economic impacts on growth and poverty reduction, there is an urgent need to initiate a major research effort that 're-visits' the methodology of cost benefit analysis (CBA) of major water investment projects. Such a research effort should also include case studies on investment paths of water infrastructure and institutions across the stages of development, hydrological characteristics, institutional contexts and policy situations. Some of these issues are discussed in detail below.

Studies on Indirect Economic Impacts of Dams under Varying Hydrological and Policy Situations

It is important that both direct and indirect impacts are captured and these impacts for different economic groups, including the poor, are quantified with the use of economy-wide multiplier models that incorporate Social Accounting Matrices (with embedded input-output analysis). Such *ex-post* analyses should be carried out for a number of large dams, medium-sized dams and small check dams under a variety of hydrological typologies (rainfall level and variability) and socio-political situations. Such case studies should be carried out both in developed countries in the United States, Europe and Australia and in developing countries such as Brazil, India, China, South Africa, Mexico, Turkey, and Pakistan.

Documenting the Costs of Inaction and Delay in Projects

It is important to document the costs of inaction or delay in the case of large multipurpose water projects that have transformed regions/

countries. Such analyses should quantify the costs of inaction in terms of adverse impacts on food security and regional incomes if a particularly large or medium-size dam had not been built or delayed considerably.

Comprehensive Benefit-Cost Analysis of Projects

Currently, in the *ex-ante* evaluation of water investment projects, the full range of costs and benefits are not assessed and evaluated. Furthermore, because these tools were developed in countries that already had a 'minimum platform' of infrastructure, these have a heavy focus on marginal analysis that may be inappropriate in some developing countries. In particular, in *ex-ante* project analysis, only direct benefits are calculated and indirect economic impacts stemming from inter-industry linkages and consumption-induced effects are not incorporated in the analysis.

It is sometimes asserted that "Shadow prices that include carefully traced indirect changes in value-added include the multiplier effects while minimising the danger of double counting. And...Most of the multiplier effect is accounted for if we shadow-price at opportunity cost" (Gittinger, 1982).

Both in principle and when applied to the stream of direct inputs and outputs associated with a project, the shadow prices used in a full cost-benefit analysis account fully for all downstream effects because they contain, among other things, all the information obtainable from a semi-input-output analysis. In practice, however, shadow prices are not usually calculated on the basis of such complete information, but are derived using the shortcut methods proposed in the standard literature. Further, it is stated[2] that "capturing multiplier effects in cost benefit analysis requires, in principle, that the estimates of the relevant parameters pertain to the region in question. Since developing countries are usually regionally heterogeneous and input-output tables are not available at the regional level, the cost-benefit analysis usually may not capture downstream effects." This was also suggested by Chopra (1972) in a pioneering study on the regional distributive effects of the irrigation

2. For details, see Bell and Devarajan, 1980.

component of the Bhakra dam. According to her, "...An application of the regional input-output technique would have been far more precise but the complete absence of data (input-output tables over time) would have necessitated far more 'heroic' assumptions than those we have had to make in adopting a modified cost-benefit analysis."

Hence, it is necessary to carry out a number of case studies of water investments using both the extended cost-benefit analysis (to include multiplier effects) and multi-sectoral models for explicit quantification of indirect economic impacts. Since, in practice, shadow prices are not based on regional semi-input output tables, indirect economic impacts have to be estimated explicitly along with the direct economic impacts in *ex-ante* evaluation of water projects. This will require assessment and quantification of such indirect impacts either by using a multiplier value based on other studies of a similar project or quantifying both direct and indirect economic impacts by using a multi-sectoral economy-wide model for the region. For large water investment projects that have region-wide effects, it may require the use of regional level computable general equilibrium (CGE) models or multiplier models based on Social Accounting Matrices. Such studies may relate to *ex-ante* evaluation of large multipurpose projects or medium-size dams under consideration in a number of developing countries.

Case Studies of Multipliers in Non-Water Investments

Since such indirect economic impacts are also significant for a number of large infrastructure projects, these impacts should be quantified for all such projects. Quite often such projects become an important mechanism of economic development and poverty alleviation in not only the region where these projects are located but also in the rest of the country. Hence, it is necessary to carry out case studies on the multiplier impacts and indirect economic impacts of non-water infrastructure investments such as highways, roads, ports, railways, power plants and so on. Such *ex-post* evaluations should be carried out both in developed and developing countries and results may be used for comparison with indirect economic impacts of dams.

References

Bell, C. and S. Devarajan (1980). "Semi-Input-Output and Shadow Prices: A Critical Note", *Oxford Bulletin of Economics and Statistics*. August.

Briscoe, John (2003). *Note on Revisiting Economic Analysis of Large Water Projects*. Draft. June.

Chopra, Kanchan Ratna (1972). *Dualism and Investment Patterns: An Analysis of Regional Contrasts*. New Delhi: Tata McGraw-Hill Publishing Company Ltd.

Gittinger, J. Price (1982). *Economic Analysis of Agricultural Projects*. Baltimore: Johns Hopkins University Press.

Part II

Case Studies

6 Bhakra Multipurpose Dam System, India

Ramesh Bhatia and R.P.S. Malik

Introduction and Overview

The Bhakra dam system in the northern part of India has contributed significantly to increases in irrigated area and the output of agricultural commodities and electricity over the last 40 years. These increases have inevitably generated downstream growth in many other sectors both in the regional economy as well as in other parts of the country. This chapter focuses on the indirect economic impacts of the Bhakra dam system and the subset of social impacts linked with changes in household expenditure, income, and its distribution. It considers as direct all economic impacts descending from the construction of the dam, the water stored and other services provided by the structure, and changes in flow regimes—regardless of whether these impacts were initially planned. Indirect and induced impacts are those that stem from the linkages between the direct consequences of the dam project with the rest of the economy. Among them are impacts due to changes in output and input use in sectors other than those affected directly by the dam, or changes in relative prices, employment and factor wages. The indirect and induced impacts have been estimated in terms of a multiplier value. As already discussed, multipliers are summary measures expressed as a ratio of the total effects (direct and indirect) of a dam project to its direct effects.

Apart from estimating multipliers, the models used in the study explicitly consider income distribution by disaggregating households into various income categories. The impact of the dam system is explicitly considered on the changes in income levels of landless workers, agricultural labour, self-employed farm households and urban households.

The material presented in this Chapter is organised as follows. We first present some salient features of the Bhakra dam system and its major outputs and impacts. This is followed by the approach and methodology used for the multiplier analysis. The estimates of multiplier values for the project multiplier are discussed next followed by a brief discussion of the multiplier effects of the Bhakra dam outside the project region. The social and environmental aspects of the Bhakra dam are presented next followed by the conclusions on the multiplier effects of the dam.

The Bhakra Dam System

The Bhakra Nangal Project, located in North-West India, is a multipurpose river valley project encompassing the three eastern flowing rivers—Sutlej, Beas and Ravi—as well as the Yamuna. The project is a splendid example not only of Integrated Water Resources Management (IWRM), as the project was planned not only on a basin level but also included inter-basin transfer of water from surplus basins to deficit basins. It is also a splendid example of inter-State cooperation between the states of Punjab, Haryana, Rajasthan and Jammu & Kashmir on sharing of water. The Bhakra Nangal Project thus, is inseparably interwoven with this grand integrated project of water management, known earlier as Bhakra-Beas-Rajasthan project, which was subsequently called the Master Plan for harnessing the waters of the Sutlej, Beas and Ravi. The Master Plan, envisaged as under;

1. The three rivers will be developed as an integrated unit and in addition will be integrated with the Yamuna river in so far as the Narwana branch of the Bhakra main line canal allows Sutlej waters to be used in the Western Yamuna canal areas.
2. Storage dam at Bhakra on the Sutlej to meet the needs of erstwhile Punjab via diversion at Nangal, Ropar and Harike with a small portion going to Rajasthan.
3. In order to fill the Bhakra reservoir in view of the heavy demands upon it from erstwhile Punjab and Rajasthan, Beas water will be diverted into the Sutlej via a high level canal and two tunnels between the Beas and Pandoh and the Bhakra reservoir (Beas-Sutlej link).

4. The remainder of the Beas water will continue to flow into the Pong reservoir to be released as required at Harike. Thus, the main burden of the Rajasthan canal component would be covered by Pong reservoir in Beas.
5. Harike also benefited from the Ravi-Beas link canal built in 1952-1954 from Madhopur headworks to Chakki tributary of the Beas. Ultimately the dam would be built upstream of Madhopur on the Ravi.

The process of integrated development has been carried out in a phased manner from 1954. As part of this process the Bhakra dam was the first to be built, which was completed in 1963. The Pong dam was completed in 1974 and the Beas-Sutlej link was completed in 1977. With the completion of the Ranjit Sagar dam recently the integrated master plan has been completed (See Figures 6.1 and 6.2).

Bhakra Dam was described by the first Prime Minister of India Jawahar Lal Nehru as the "temple of resurgent India." The construction of the Bhakra dam system, resulted in a very significant increase in the gross irrigated area, output of agricultural commodities and generation of hydro-electricity. In addition to these direct economic impacts, the Bhakra system induced several indirect economic impacts. These direct and indirect economic impacts transformed the economy of the Punjab-Haryana region resulting in a significant reduction in the incidence of poverty in this part of the country. Higher output and consequently a higher level of marketable surplus of foodgrains in the Punjab-Haryana region made it possible to provide these foodgrains to the urban poor at affordable prices even in far away regions of the country. Increased demand for labour in agricultural and non-agricultural activities provided jobs and substantially higher incomes to migrant workers from such far off places as Bihar and Uttar Pradesh. The remittances by these migrant workers to their families back home helped contribute to the development of poorer regions of the country.

The Bhakra dam, completed in 1963, is a 225.55-meter (740 feet) high straight concrete dam. The lake created by the dam is 162.48 square km in area with a gross storage of 9340 million cum. The Nangal Dam, situated about 13 kms downstream of the Bhakra dam, is 29 metres (95 feet) high and comprises 26 bays of 9.14 metres (30 feet)

Figure 6.1

The Master Plan for Development of Water Resources

A. Satluj Beas Ravi Development Master Plan

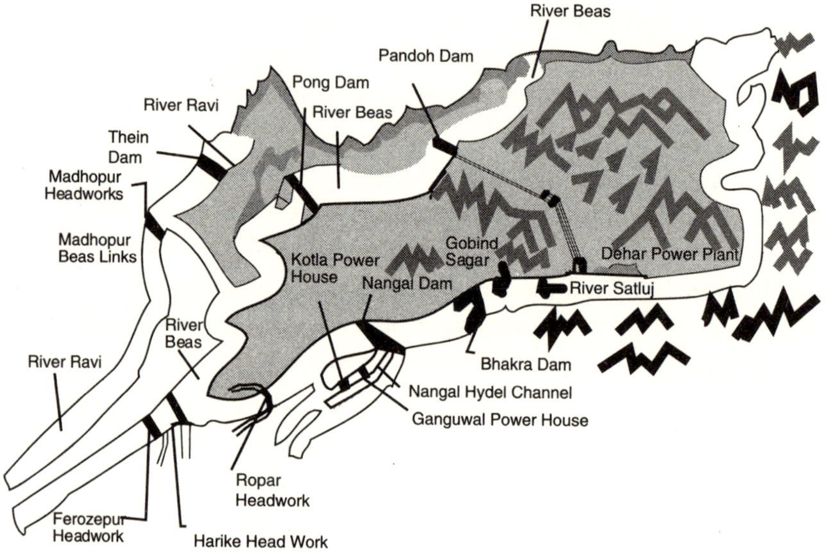

each. It is designed to pass a flood of 10000 cumecs (350000 cusecs). The dam diverts the water of the river Sutlej into the Nangal Hydel Channel and Anandpur Sahib Hydel Channel for power generation and irrigation purposes. The Nangal pond acts as a balancing reservoir to smoothen out the diurnal variation in releases from the Bhakra Power Plants. The Nangal barrage provides the headwork for the 414 cumecs Nangal hydel channel with hydro stations at two falls at Ganguwal and Kotla with a combined installed capacity of 154 MW. The Bhakra main canal (360 cumecs) takes off from the Ropar headwork, 60 kms from Nangal Dam to irrigate large tracts and firm up supplies to some older systems. The Beas Project I comprises a diversion dam at Pandoh, several hydel channels, tunnels, control works, a balancing reservoir and Dehar power plant while the Beas Project unit II comprises the Pong dam, tunnels, spillways and Pong power plant. The salient features of the Bhakra dam system are given in Table 6.1.

The Bhakra dam and its various components were constructed at different stages over a number of years. At prices prevailing in the early

B. River Sutlej, Beas, Ravi and Connected Main Canals

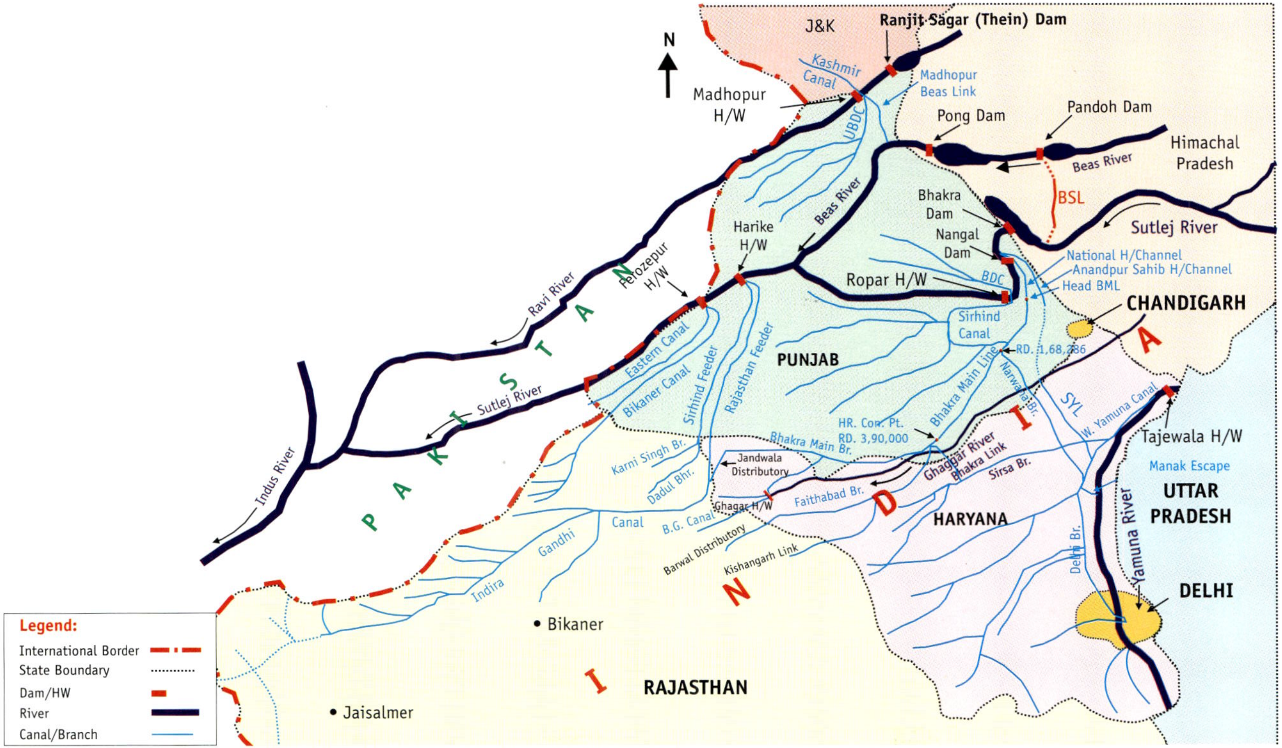

Figure 6.2

The Bhakra-Beas System

1950s, the estimated cost of the Bhakra and Nangal dams with canal systems was Rs. 2426 million (Table 6.2). At the time of its construction in 1967-68, the Bhakra right bank power house cost Rs. 983 million. The upgradation of the turbines in 1980-81 to add 150 MW capacity involved an additional expenditure of Rs. 100 million. A sum of Rs. 4492 million was spent on the construction of the Beas-Sutlej link project in 1976-77 while the construction of the Pong Dam cost another Rs. 3259 million. This totals to not more than about Rs. 11260 million at prices prevailing in those years when these investments were made (Table 6.2).[1] The capital cost of the entire Bhakra system at current prices (2001-02) is estimated at Rs. 123,478 million or US$ 2,680 million.

1. These are sum of undiscounted costs at the then prevailing prices.

Table 6.1

Bhakra-Nangal-Beas Project: Salient Features

	Bhakra Dam	*Nangal Dam*	*Beas Unit I-Pandoh Dam-DeharPower Plant*	*Beas Unit II*
Type of Dam	Concrete straight gravity		Earth cum Rockfill	Earth Core Gravel Shell
Height above Deepest Foundation	225.55 metres		76.2 metres	132.59 metres
Height above River Bed	167.64 metres	29 metres	61 metres	
Length at Top	518.16 metres	304.8 metres	255 metres	
Width at Top	9.14 metres		12.19 metres	
Catchment Area of Reservoir	56980 square Kms			12560 square metres
Area of Reservoir	162.48 square Kms			
Live Storage Capacity	6911 million cum		18.56 million cum	7290 million cum
Gross Storage Capacity	9340 million cum		41.00 million cum	8570 million cum
Number of Power Houses/ Units	2	2	6	6
Capacity of Power Plants	1325 MW (5×180+5×157)	154 MW (4×24+2×29)	990 MW (6×165)	396 MW (6×66)

Table 6.2

Capital Costs of the Bhakra System Components at Current Prices and 2001-02 Prices (Rs. Millions)

	Original Cost (Rs Million)	*Year of Investment*	*In 1993-94 at 1993-94 Prices*	*In 2001-02 at Current Prices*
Bhakra and Nangal Dams	2426	1957-58	33671	55751
Power House	983	1967-68	7404	12259
Upgradation of power house	100	1980-81	308	510
Beas-Sutlej Link	4492	1976-77	19236	31850
Pong Dam	3259	1976-77	13956	23108
Total	11260		74575	123478

Source: Bhakra Beas Management Board and Author's Calculations.

Keeping in view the complex system of water management, sharing and transfers within the system, of which Bhakra-Nangal project is the most important component, it will not be prudent to confine the benefits flowing from the Bhakra-Nangal project to the boundaries as defined in the original project reports and documents when this project was planned. Since the Bhakra-Nangal is the core around which all the activities described above have been carried out in phases, we have therefore interchangeably used the Bhakra system and the Bhakra-Nangal project to reflect some of these inter-linkages. Hence, the estimates of irrigated area in different states and the output of electricity reflect the benefits from all the components of the larger Bhakra-Nangal-Beas system termed here as the Bhakra system. These figures will be different from those given in the official documents that refer mainly to the Bhakra dam.

The Major Outputs of the Bhakra Dam System

The major outputs of the Bhakra dam system, *inter alia*, have been:

(i) Increased availability of water for irrigation resulting in higher agricultural output

(ii) Availability of water for industry, household enterprises and for households and livestock

(iii) Generation of hydro power, and

(iv) Moderation of floods reducing flood damage significantly.

The construction of the Bhakra system has made available huge amounts of water for meeting irrigation requirements of the region. The command area of the system—covers the three States—Punjab, Haryana and Rajasthan. Delhi also gets a small share of 1 per cent of the system water for meeting partially its drinking water requirements (Figure 6.3). (Gopalakrishnan, 2000).

A number of studies have been undertaken in the past to estimate the direct economic impacts of the Bhakra dam. In his pioneering study on the cost benefit analysis of a water investment project, Raj (1960) attempted to break down costs and returns of the project into certain categories and evaluated these according to selected criteria so as to determine the social economic value of the project as distinguished from its private economic value. According to the author, "the decision to

Figure 6.3

Shares of Different States in Water from the Bhakra System

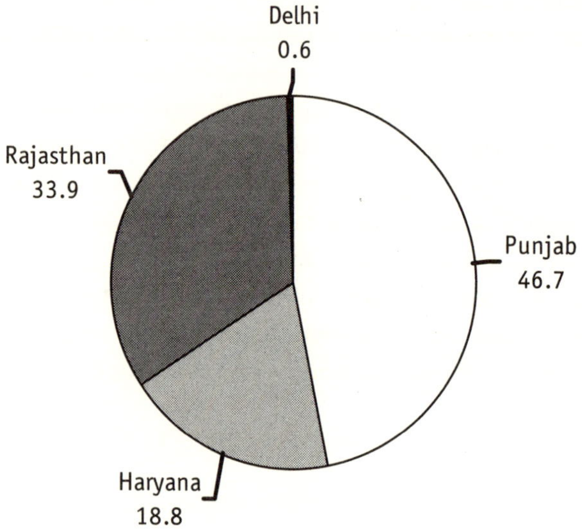

Source: Prepared by authors from Gopalakrishna (2000).

locate a fertiliser plant at Nangal appears to have been, by itself, a correct one. The production of nitrogenous fertilisers is not only dependent on the availability of cheap power but it is vital to the economy of this region; in fact, it would make possible more effective utilisation of the irrigation facilities provided by the project" (Raj, 1960: 129).

Chopra (1972) presents results on the regional distributive effects of the irrigation component of the Bhakra dam using Social Cost Benefit approach. An evaluation of the investment in irrigation has been undertaken under alternative assumptions regarding investment costs and a number of alternative values of the parameters such as the shadow price of labour, the shadow price of foreign exchange and the premium on funds devoted to investment. The results show that regional incomes have increased to a far greater extent than aggregative incomes. The empirical analysis also shows that in the case of aggregative effect, the elements that are relatively more significant are the foreign exchange component of costs and the shadow price of foreign exchange. Interdependence between sensitivity with respect to different parameters highlights the need for determining them as part of the planning process. According to Verghese (1994), the Bhakra-Beas complex project "has been an outstanding success. It has built India and has long paid for itself many times over." The Bhakra-Beas projects have brought agricultural stability to Punjab and Haryana. "In the bad drought year of 1987, when many parts of the country reeled under distress, farm production in northwest India was unaffected. Thanks to that, India could face and overcome the crisis. Employment and income have risen appreciably and Punjab has attracted migrant labour for sowing and harvesting from as far as Bihar and Orissa" (Verghese, 1994: 39-40)

Additional Irrigation and Agricultural Output

The Bhakra system irrigation command is spread over several districts of the States of Punjab, Haryana and Rajasthan. Almost 86 per cent of the geographical area in the command area in Punjab and Haryana is under cultivation. More than 73 per cent of the gross cropped area in the region is devoted to the cultivation of food grains.

The availability of irrigation and consequent shifts in the cropping pattern have been accompanied by adoption of yield increasing technological innovations. Almost 96 per cent of the area under wheat and 86 per cent of the area under paddy-rice is sown with HYV seeds. The methods of cultivation have also undergone significant changes over the years. Large-scale mechanisation of such farm operations as land preparation, sowing (for wheat), irrigation, harvesting and threshing have made the region agriculturally the most advanced region in the country. Most of the area under rice and wheat is now harvested mechanically using combine harvesters.

The availability of irrigation water through this dam-canal network in addition has helped in increased recharge of groundwater. The water that is lost as seepage in canal irrigation system is available as surface return flow or groundwater recharge. Irrigation by groundwater on a very large area in this region, with an average rainfall of about 60 cm only, has largely been made possible through recharging of the groundwater by seepage through canals. It has been estimated that the natural recharge to groundwater in Punjab can sustain hardly half the existing number of private tubewells in the State. In other words, the availability of canal network has enhanced groundwater availability by a factor of two (Dhawan, 1993).

Adoption of the complementary yield enhancing technological innovations consequent upon availability of assured water supply has led to a very significant jump in crop yields. The yields of rice and wheat in the region are higher than yields obtainable in any other region of the country and compare favourably with the yield levels obtaining in some of the advanced countries of the world.

The availability of year round irrigation water through the dam-canal network (Figure 6.4) and groundwater pumping has helped in bringing large tracts of cultivated area under irrigation. Additional gross irrigated area has been of the order of 7.1 million hectares over a period of 40 years (Table 6.3). The total foodgrain production in the Bhakra command area during the year 1996-97 was of the order of 27 million tonnes, an additional output of 25.2 million tonnes compared to the food output in the mid 1950s.

It may be emphasised that in a low-rainfall (less than 600 mm per year) area such as the Bhakra command, adequate and timely irrigation is an essential input without which the high-yielding variety seeds cannot be used. Further, the level of use of chemical fertilisers also critically depends on the availability of reliable irrigation water. It is in this context that irrigation from the Bhakra dam and from groundwater pumping in the area has been a 'leading input' without which it would not have been possible to attain such high crop yields in rice, wheat and cotton as have been obtained in Punjab and Haryana.

Impact on Food Security of India

The availability of huge surpluses of foodgrains from the region has significantly reduced the dependence of the country on imports of foodgrains for meeting the foodgrain requirement of the population, insulated the country to a large extent against droughts, made Indian agriculture more sustainable, contributed to the food security of the country and has helped in reducing wide fluctuations in the prices of foodgrains.

Figure 6.4

Water Releases from Bhakra During Different Months, 1998-99 (cumec days)

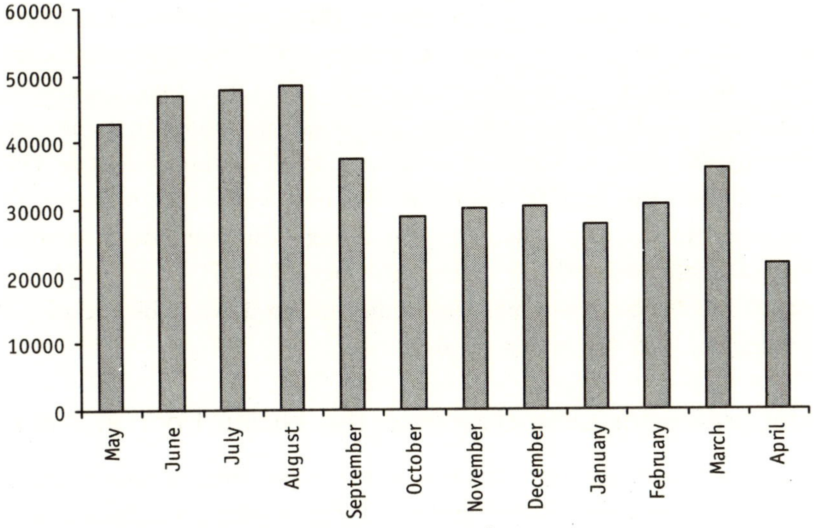

Source: Based on data from *www.bbmb.gov.in*

Table 6.3

Irrigated Area and Production of Foodgrains in Bhakra Dam System in Punjab and Haryana: 1955-56 and 1996-97

	1955-56 Without Project	*1996-97 With Project*	*Per cent Increase/Change (1996-97 over 1955-56)*
Gross Cropped Area (Million Ha)	7.0	11.2	60
Net Irrigated Area	3.0	5.6	87
Gross Irrigated Area	3.2	10.3	222
Production of Rice and Wheat (Million Tonnes)	1.8	27.0	1400
Fertiliser Consumption (Mn Nutrient Tonnes)	Neg	1.5	

Source: Statistical Abstracts of Punjab and Haryana, Various years.

Increase in surface irrigation from the Bhakra dam and resulting groundwater pumping have led to significant increases in foodgrain production. By 1980, the production of foodgrains in both states had reached around 18 million tonnes or 19 per cent of the total all-India production. As much as 55 per cent of the total increase in foodgrain production in the country over two decades (1980 over 1961) came from the Bhakra dam command. As a result of such a high increase in domestic production, the net imports of foodgrains declined substantially from 10.3 million tonnes in 1966 to almost zero in 1972. Even though the imports declined, the net availability of foodgrains continued to increase. Further, even though the population increased from 442 to 666 million during the period 1961 to 1979 the net availability of foodgrains per capita did not go down substantially and remained around 470 grams per day.

Foodgrains for Distribution to the Urban Poor in India

The huge production of foodgrains in the region created vast quantities of marketed surplus of these grains. The procurement agencies of the Central government every year purchases enormous amounts of this marketed surplus of foodgrains for maintaining the national buffer stock of foodgrains and running the Public Distribution System (PDS) of the country through a distribution network of hundreds of ration shops spread all over the country. During the year 2001-02 the States of Punjab and Haryana contributed 17 million tonnes of wheat out of a total of 21 million tonnes procured by these agencies from all over the country. Similarly, during the year 2000-01 out of a total procurement of 19 million tonnes of rice from the entire country, Punjab contributed 7 million tonnes while Haryana contributed another 6 million tonnes.

Impact on Hydropower Availability

The hydro power stations installed in the Bhakra system have a combined generating capacity of 2880 MW that currently generate about 14000 million Units (kWh) of electricity in a year. It has set up transmission lines running in to 3738 ckt kilometres. The power generated, benefits seven states/union territories. Of the total of 13796 million units of electricity transmitted during 1998-99, about 38 per cent was utilised by Punjab, 27 per cent by Haryana and 23 per cent by Rajasthan (Figure 6.5). The generation of hydropower has enabled the region to more than achieve the desired hydro-thermal mix of 40:60. As of March 1996, Punjab had an installed power generating capacity of 3509 MW of which about 51 per cent was hydel power, mainly from the Bhakra system. In Haryana, of the total installed generation capacity of 1762 MW, hydro constitutes about 50 per cent. In comparison at the all-India level the share of hydropower in the total electricity generating capacity was only 25 per cent.

Bhakra-Beas Management Board (BBMB) supplies about 50 per cent of the total hydropower in the Northern region besides meeting the peak demand of about 2500 MW of the Northern Grid. The availability of hydropower has helped BBMB generate huge revenue. The 14110 million units of hydropower generated in 1998-99 evaluated at a conservative price of Rs. 2 per unit is worth Rs. 28,220 million (Table 6.4). The operational cost of power generation and transmission work out to less than Rs. 0.10, perhaps the cheapest in the world (Gopalakrishnan, 2000).

Figure 6.5

Share of Different States in Electricity Generated in the Bhakra System

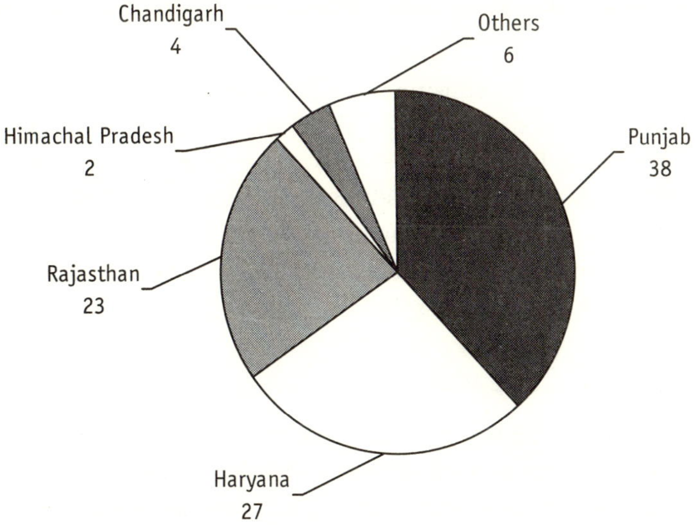

Source: Based on data from *www.bbmb.gov.in* and Author's Calculations.

Table 6.4

Total Generation and Value of Electricity from the Bhakra System

Year	*Total Generation (Million Units)*	*Value @Rs. 2 per Unit (Rs. in Million)*
1994-95	12223	24446
1995-96	12086	24172
1996-97	12085	24170
1997-98	10607	21214
1998-99	14110	28220

Source: Based on data from *www.bbmb.gov.in* and Author's Calculations.

Rural Electrification

As a result of availability of power from the dam, 100 per cent electrification was achieved in Punjab and Haryana by the year 1976. Most of the towns and villages in Western Rajasthan have also been electrified. In comparison, at the all-India level average rural electrification was around 40 per cent. In other states, the percentage of villages electrified ranged from 28 per cent in Bihar, 32 per cent in Uttar Pradesh and 52 per cent in Gujarat. Thus, the Bhakra system provided electricity to all the villages and towns in the early seventies resulting in significant improvements in education, health, and quality of life. Village electrification also helped in the development of rural industries and agro-based industries in these areas resulting in all-round prosperity and poverty reduction.

The overall power consumption per 1000 population as also per square kilometres of area amongst all States in the country (except the highly urbanised Delhi) is the highest in Punjab. The electricity consumption (for utilities) per 1000 population during 1996-97 was 776 kWh while consumption per square kilometre area was 347,619 kWh. The corresponding figures for all-India average during the year were 297 kWh and 85,240 kWh respectively.

Installation of Tubewells

The increased availability of power enabled farmers to install private shallow tubewells for groundwater withdrawals. This has already been discussed above. These tubewells during the year consumed about 8700 million kWh of electricity. The proportion of electricity consumption for irrigation pumping to total electricity consumed in the region worked out to about 40 per cent.

Impact on Reduction in Poverty in Punjab and Haryana

The availability of irrigation water and electricity for households, agriculture and industry and the consequent growth in rural and urban areas that followed has helped in very significantly bringing down poverty in the two states of Punjab and Haryana. As already mentioned the benefits of this economic development in the region have been shared both by rural as well as urban areas. As a result this growth scenario has led to a steep decline in poverty in the rural areas of both the States—this despite the fact that a very large number of labourers from poor regions of the country having migrated to this region. The poverty in the two States is much lower in comparison to both all-India figures as also in comparison to other major States of the country. In rural areas, as shown in Table 6.5, the percentage of people below the poverty line during 1999-2000 was only 6.35 per cent in the Punjab and 8.27 per cent in Haryana as compared with 27 per cent in India. In urban areas, the per cent of people below the poverty line was only 5.7 in Punjab and 10 per cent in Haryana compared to 23.6 in the country as a whole. The Bhakra dam has played a very significant role in the last about 40 years in bringing about significant reductions in poverty in the states of Punjab and Haryana. It is expected that the remaining command area under the Bhakra system also experienced similar poverty reduction impacts.

Table 6.5

Poverty Ratios: Per cent of People Below Poverty Line in Punjab, Haryana and All-India

State	*Rural*		*Urban*		*Rural+Urban*	
	1973-74	*1999-2000*	*1973-74*	*1999-2000*	*1973-74*	*1999-2000*
Punjab	28.21	6.35	27.96	5.75	28.15	6.16
Haryana	34.23	8.27	40.18	10.00	35.36	8.74
All-India	56.44	27.09	49.01	23.62	54.88	26.10

Source: Economic Survey, 2001-02.

Flood Control Benefits of the Bhakra System

One of the most reliable structural methods to physically control floods is to store the excess water in major dams. The Bhakra and Pong dams were not planned as flood control reservoirs though these were expected to provide a cushion against floods. However, the Bhakra and Beas dams have helped in the moderation of floods by absorbing peak flood inflows as would be clear from the data in Table 6.6.

Table 6. 6

Impact on Moderation of Floods

Year	Bhakra Dam		Beas Dam	
	Peak Flood Inflow (Cumecs)	*Max Released (Cumecs)*	*Peak Flood Inflow (Cumecs)*	*Max Released (Cumecs)*
1978	10719	3887	—	—
1988	9004	4209	10692	7972
1992	6625	1542	6587	2741
1994	6364	1695	8931	1406
1995	8977	1658	9623	5633
1998	5244	1330	4270	—

Source: Gopalakrishna (2000).

The Bhakra dam has solved the problem of recurring floods in the Sutlej River, where a large extent of the wide bed of the Sutlej below the Bhakra dam is now utilised for intensive agriculture. (Gopalakrishnan, 2000).

Benefits from Integrated Management of Hydropower, Surface Irrigation and Groundwater Use

Benefits from the integrated management of hydro power and irrigation have been analysed in a few studies on the Bhakra system and the integrated operation of the Beas-Sutlej system (Minhas *et al.*, 1972; Rao and Ramaseshan, 1985; Rao and Ramaseshan, 1985a). In the Bhakra system, (the then) planned levels of power generation vary widely between 766 MW from December through April to 1697 MW in September. Conjunctive utilisation models have been developed for integrated management of surface and groundwater irrigation. A linear programming model of the system was developed (Rao and Ramaseshan,

1985a) that attempted to maximise the level of firm power and satisfied irrigation demands in each of the sub periods. The power required to lift groundwater was over and above the firm power that was to be supplied. This integrated framework led to a better understanding of the interaction between the irrigation and firm power objectives. The results show[2] that conjunctive utilisation can increase firm power by at least 36 MW (over and above a firm power of 797 MW). The results also indicate, contrary to the then practice, groundwater use occurs generally between May and November and not from December to April. The firm power level reached in a dry year is 200 MW (i.e. 20 per cent) less than that reached in a dependable year. The levels of irrigation and power planned for a dependable year from the reservoirs of the Beas-Sutlej system can be attained even in a dry year by conjunctive utilisation of surface and groundwater.

Multiplier Effects of the Bhakra Dam System in the Punjab State

As discussed above, the Bhakra dam system has benefited not only the states of Punjab, Haryana, Rajasthan, Himachal Pradesh and Delhi but also the entire country. For the estimation of the multiplier effects of the dam for the present study, we will however confine to the direct and indirect benefits of the dam system in only a part of the benefited area *viz.* the state of Punjab, the biggest beneficiary of the dam system. Further, the analysis captures the main effects of the dam system in a typical year during its lifetime *viz.* 1979-1980, approximately 20 years after the construction of the Bhakra dam was completed. By 1979-80 the benefits from the other components of the Bhakra system were also available in the Punjab state. The choice of geographical area *viz.* Punjab state and the choice of the year 1979-80 for the present analysis has in large part been dictated by the availability of an Input-Output (I-O) table (for constructing the Social Accounting Matrix) for the state of Punjab (Bhalla *et al.*, 1990) and not for the entire Bhakra system. In India I-O tables at the state level are available only for a few states and that too for only one or two points in time. It is however, expected that similar

2. It is likely that the Bhakra-Beas Management Board is operating the system so as to optimise the power and irrigation output.

indirect economic impacts are also available in other areas in Haryana and Rajasthan that have benefited from the Bhakra dam system.

For the last four decades or so, the Bhakra dam system has played an important role in the provisioning of irrigation and electricity to the Punjab state. During the period between 1955-1956 to 1979-80, the availability of net irrigated area has increased by almost 55 per cent while the gross area irrigated has increased by 125 per cent (Table 6.7). During the year 1979-80, the Bhakra system command produced about 3 million tonnes of rice and 7.4 million tonnes of wheat besides a host of other crops including sugarcane, cotton etc. (Table 6.8 and Figure 6.6).

The Approach and Methodology

Major outputs of the Bhakra multi-purpose dam system, like any other large multipurpose dam, and reservoir project include: hydropower, irrigated agriculture, water supply, fishing, flood control, drought prevention and value of recreational activities/tourism revenues. In

Table 6.7

Estimated Irrigated Area in the Bhakra System Command in Punjab: 1955-56 and 1979-80 ('000 Ha)

	1955-56	1979-80
Net Irrigated Area (NIA)		
- By Canals	1340	1478
- By Wells, TW and Other Sources	852	1913
Total NIA	2192	3391
Total Gross Irrigated Area (GIA)	2426	5448

Source: Statistical Abstract of Punjab, Various Issues.

Table 6.8

Area and Production of Important Crops in the Bhakra System Command: 1979-80

Crop	*Area (000 ha)*	*Production ('000 MT)*
Rice	1110	2923
Wheat	2787	7362
Sugarcane	59	294
Cotton	626	2045

Source: Statistical Abstract of Punjab and Haryana, Various Issues.

Figure 6.6

Area and Production of Important Crops in Bhakra System Command: 1979-80

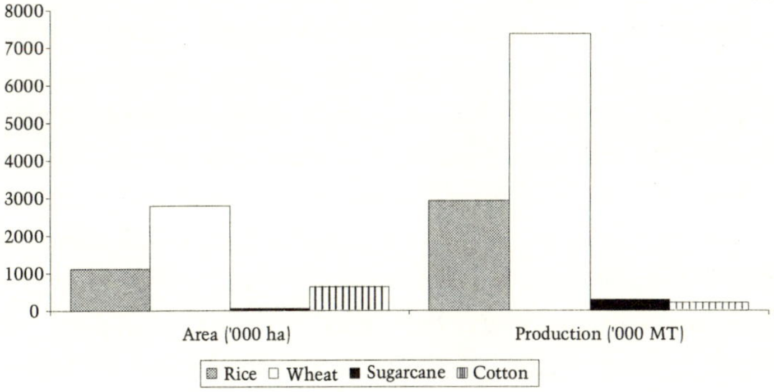

Source: Table 6.8.

addition, there are non-irrigation benefits of irrigation canals in terms of water for households, livestock, household enterprises, animals and birds, among others. The negative direct impacts of the dam project include, inter-alia, social, economic, and cultural costs of resettlement; value of lost ecosystem/historical/cultural heritage sites; opportunity cost of lost output from inundated area and reduction in fish output upstream. For the purpose of this paper, direct costs and benefits are all those that descend from the construction of the dam, as well as economic benefits accruing to all beneficiaries of the water stored and other services provided by the structure and the economic impacts—positive or negative—to all populations, species and ecosystems affected either by the construction or by the changed regime of the river.

Indirect and induced impacts are those that stem from the linkages between the economic, environmental, social or other direct consequences of a dam project with the rest of the economy, as well as the non-economic impacts descending from such linkages. Among them are impacts due to changes in output and input use in sectors other than those affected directly by the dam, or changes in relative prices, employment and factor wages.

Major outputs of the dam have significant inter-industry linkages and result in an increase in the demand for output of other sectors of the economy. Hydropower produced from a multipurpose dam provides electricity for households in urban and rural areas and for the increased output of industrial products (e.g., fertilisers, chemicals, machinery etc.). Changes in the output of these industrial commodities require inputs from other sectors such as steel, energy, and chemicals, among others. Similarly, water released from a multipurpose dam provides irrigation for the increased output of agricultural commodities. Changes in the output of these commodities require inputs from other sectors such as energy, seeds, fertilisers, etc. Further, increased output of some agricultural commodities encourages setting up of food processing and other industrial units. Thus, both the increased output of electricity and irrigation from a dam result in significant backward linkages (i.e., demand for higher input supplies) and forward linkages (i.e., providing inputs for further processing).

Increased outputs of industrial and agricultural commodities generate additional wages and incomes for households. Higher incomes result in higher consumption of goods and services that, in turn, encourages the production of various agricultural and industrial commodities. Further, changes in output generated by the project may affect prices of direct project outputs, inputs, substitutes, complements and factors of production. Changes in wages and prices have both income and substitution effects on expenditure and saving decisions of different owners of factors, which further impacts the demand for outputs both within the region and throughout the economy. Induced impacts reflect the feedbacks associated with these income and expenditure effects, and also include any impacts of changes in government revenues and expenditures that result from the project.

Major outputs from a dam thus generate both inter-industry linkage impacts and consumption-induced impacts on the regional/ national economy. The level of indirect impacts of a dam on regional output and value-added will depend on the strength of linkages among various sectors of the economy. Multiplier analysis is one of the approaches for quantifying the magnitude of inter-industry linkages and consumption-induced effects, relative to purely direct impacts.

Multipliers are summary measures that reflect the total effects of a project in relation to its direct effects. A multiplier of 1.90, for instance shows that for every one dollar of value-added generated directly by the project at maturity, another 90 cents were generated in the form of indirect or downstream effects. Thus, a multiplier is a ratio of the total effects (direct and indirect) of a dam project to its direct effects.

Estimation of multipliers requires careful analysis of the direct effects of a dam. This involves the quantification and valuation of major outputs of the dam and the assessment of the share of direct effects that is attributable to the dam project. Valuation of outputs/benefits from a water project is quite a complex matter, particularly in the case of multi purpose dams.

As already discussed, the multiplier analysis reported in this chapter is confined to the state of Punjab, a major beneficiary of the Bhakra dam system, and the analysis has been done for a year (1979-80) for which adequate data were available from a detailed study (Bhalla *et al.*, 1990). The study employs a SAM-based multiplier model to estimate project multipliers using a Social Accounting Matrix for Punjab.[3] The model has been used to compute the values of relevant variables in the 'With Project' situation with their counterparts in the hypothetical case that the project had not been undertaken. This set of variables comprises all the elements of a SAM for the region in each situation, assuming fixed prices. In measuring the impact of the project, an attempt is made to assess the situation in the region (the Punjab state) for the hypothetical case of 1979-80 in the absence of the project. This has been done by assuming that all autonomous changes would have taken place except the effects of changes due to major outputs of the project, namely irrigation and hydro-electricity. This hypothetical case in the absence of the project[4] is termed as 'Without Project' scenario for 1979-80.

---

3. The approach and the methodology used in estimating the multiplier effects of the Bhakra dam are similar to those adopted in studies on the Muda dam in Malaysia (Bell, Hazell and Slade, 1982) and indirect effects of the green revolution in south India (Hazell and Ramaswamy, 1991).

4. Unlike in the pioneering study of the Muda dam (Bell *et al.*, 1982), it was not possible to prepare a SAM for the pre-project situation because not even input-output tables were available during 1952-53 when most of the additional irrigation was not available from the Bhakra dam.

As shown in Table 6.9, for estimating a project multiplier value for the Bhakra dam, for the numerator, we need to estimate the regional value-added (for Punjab state) under 'With Project' situation as well as the regional value-added under 'Without Project' situation. For the denominator, we need to estimate the value-added from the sectors that are directly affected by the major outputs of the dam (namely agricultural output, hydro electricity, water supply etc.).

Table 6.9

Values of Variables Required for the Estimation of a Project Multiplier of the Bhakra Dam Project, India

Definition of Project Multiplier	Regional value-added with project	-	Regional value-added without project
	Value-added of agriculture and electricity with project	-	Value-added of agriculture and electricity without project

Regional value-added under 'With Project' situation has been estimated by using the SAM-based multiplier model. The model has been used to estimate the SAM coefficients for each household category and for each production sector. These SAM coefficients have then been used to estimate the regional value-added under 'Without Project' situation by fixing the outputs of sectors directly affected by major outputs of the dam, namely irrigation and hydropower. Value-added of sectors directly affected by the project have also been estimated from the SAM model. The agricultural sectors directly affected by the project for which the outputs have been fixed are: wheat, rice, sugarcane, cotton, gram and other agriculture. The values of agricultural output and electricity sectors under the 'Without Project' situation have been estimated under a variety of assumptions regarding availability of canal irrigation, groundwater pumping and hydro power and the technology choices affecting area and yield of major crops. The details of the assumptions made in estimating the outputs of these agricultural sectors and electricity under a 'Without Project' situation have been described in the following sections. The multiplier values for the Bhakra dam, under different scenarios, have been estimated by using the values of regional value-added under 'Without Project' situation and comparing these with the regional value-added under 'With Project' situation.

A Social Accounting Matrix (SAM) for the Punjab

This study uses an estimated social accounting matrix (SAM) to provide a detailed quantitative description of the Punjab economy in 1979-80, almost 20 years after the construction of the Bhakra dam was completed. The SAM framework provides a consistent, comprehensive, and detailed picture of the transactions in the economy (Figure 6.7). Production activities, government and households are considered and the pattern in which incomes are distributed takes its place alongside the sources of its generation. The SAM also provides a basis for the construction of a model of the regional economy that is used to estimate the direct and indirect effects of the Bhakra dam. The data source for estimating the full[5] SAM is a detailed study of the Punjab economy reported in Bhalla *et al.* (1990).

The key features of the regional economy[6] of the Punjab are analysed with the help of an aggregated SAM in Table 6.10. In this SAM,

Figure 6.7

SAM Interactions and Circular Flow of Income

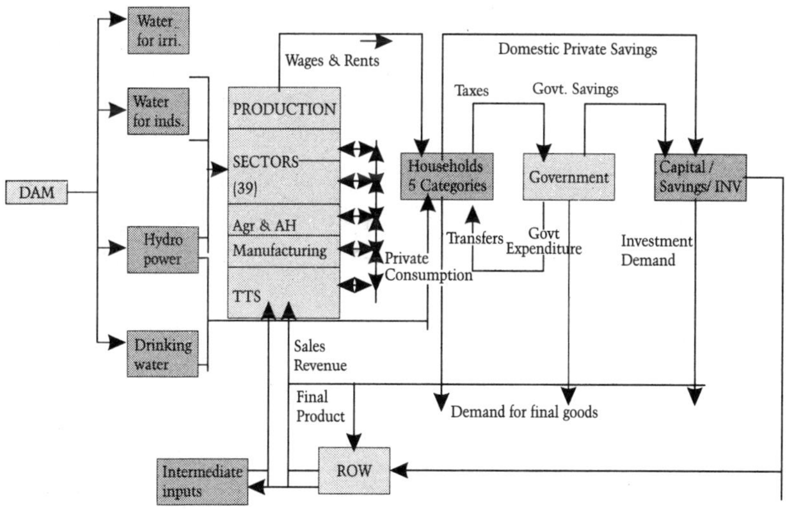

5. The full SAM of the Punjab economy is available with the authors.
6. Figures used in the SAM are marginally different from those in the data sources in Bhalla *et al.* (1990) due to balancing of figures in the SAM.

Table 6.10

Aggregated SAM for Bhakra 1979-80 (Million Rs.)

	Self-Empl.	*Agri. Lab.*	*Non. Ag. Lab.*	*Other-Rural*	*HHU*	*Agri.*	*Manf.*	*TTS*	*BIS*	*PS*	*Capital*	*Govt.*	*ROW*	*Total*
Self-Empl.	0	0	0	0	0	9562	2870	179	214	221	0	0	-540	12505
Agri. Lab.	0	0	0	0	0	5010	0	0	0	0	0	0	-1006	4005
Non. Ag. Lb	0	0	0	0	0	1389	373	38	30	28	0	0	-732	1125
Other-Rural	0	0	0	0	0	3035	2498	216	194	191	0	0	2279	8413
HHU	0	0	0	0	0	1000	3092	8217	438	3584	0	0	0	16331
Agri.	3656	1390	393	2323	2048	2239	4800	0	0	202	235	25	12737	30048
Manf.	3833	1407	424	2422	3482	3609	13397	837	12	1327	9275	181	-3245	36961
TTS	1560	560	162	1115	1000	573	4055	1431	34	773	324	79	2175	13841
BIS	55	20	6	39	48	98	380	319	83	48	0	0	29	1123
PS	472	169	49	337	691	0	0	971	18	347	0	0	4266	7320
CapitaL	2296	235	26	1405	7939	749	1354	593	34	333	0	20	0	14985
Govt.	473	165	48	666	677	232	467	414	4	246	627	0	-3636	385
ROW	160	59	17	104	445	2552	3674	627	61	22	4524	79	330	12655
Total	12505	4005	1125	8413	16331	30048	36961	13841	1123	7320	14985	385	12655	0

39 production sectors have been aggregated to five categories. In 1979-1980, the gross value of output in the Punjab was Indian Rupees[7] (Rs.) 89.3 billion and the value-added was Rs. 42.4 billion (47.5 per cent of the gross value of output). Agriculture accounted for 33.7 per cent of the total value of output while manufacturing accounted for another 41.4 per cent of the output. Shares of value-added in the total value-added were: Agriculture and animal husbandry 47.2 per cent; Manufacturing 20.8 per cent and construction and services 32 per cent. The net transactions with the rest of the world (ROW) were Rs. 12.66 billion. The capital accounts were of the order of Rs. 14.99 billion.

A SAM-based Multiplier Model of the Punjab Economy

A SAM-based multiplier model has been used to provide a quantitative analysis of the direct and indirect impacts of increased agricultural production and electricity output available from the Bhakra dam. The model is calibrated for the year 1979-80 for Punjab using the Social Accounting Matrix (SAM) described in the previous section.

Since the SAM is a double-entry accounting system, we can use either the row or column accounts for its presentation. The rows provide the statement of receipts for each account, whereas the columns provide the statement of expenditure. For example, the row for the k th type household represents the receipts of household k, i.e.

Total income of household k = Value-added received by household k from the production sectors + Income transfers from abroad to the household.

The corresponding column for the household provides the details of the expenditures of the household, i.e.

Total expenditure = Expenditure on domestic goods + Savings + Taxes + Expenditure on imported goods.

Table 6.11 shows the interactions among different accounts in the SAM. These interrelationships are explained in more detail in the following equations.

---

7. The current (December 2002) conversion rate is Rs. 48.4 for a US$. The exchange rate during 1979-80 was approximately Rs 10 to a US dollar. In this report we have used Indian Rupees (Rs.) for value figures using 1979-80 prices.

Table 6.11

A SAM-based Multiplier Model of the Punjab Regional Economy

Expenditures *Receipts*	*Households (HHs) k=1,..,5*	*Production Sectors i=1,2,...39*	*Capital Account*	*Government*	*Rest of the World*	*TOTAL*
Households k=1,..,5	0	$\sum_i w_{ki} X_i$	0	0	R_k	Y_k
Production Sectors i=1,2,...39	$\sum_k C_{ik} Y_k$	$\sum_j a_{ij} X_j$	J_i	G_i	E_i	X_i
Capital Account	S_k	S_i	0	S_g	S_R	F
Government	$\sum_k t_k Y_k$	$\sum_i t_i X_i$	0	0	R_g	G
Rest of the World	M_k	M_i	0	M_g	0	M
TOTAL	Y_k	X_i	F	G	M	

In the SAM-based multiplier model there are 39 production sectors including the agricultural sectors, manufacturing, electricity, gas and water supply, trade and transport services etc. (see Table 6.16 for a listing of these sectors). There are 5 types of households namely, self-employed rural households (HRSE), rural households employed as agricultural labour (HRAG), rural non-agricultural labour households (HRNA), other rural households (HRO) and urban households (HHU). The notation used in the following is as follows:

The subscript k = 1,2,...5 denotes the type of households, while i, j denote the production sectors.

Y_k = total income of households of type k

X_i = value of gross output of sector i

w_{ki} = coefficient (ratio) of value-added received by household type k to output of sector i

R_k = income transfers from abroad to households of type k

C_{ik} = proportion of income household k that is spent on the purchases of sector i's output

a_{ij} = input-output coefficient for production sectors

J_i = investment demand of sector i 's output

G_i = government purchases of sector i 's output

E_i = exports of sector i 's output
G = total tax revenues
t_k = tax rate for households of type k
t_i = tax rate for production sector i
R_g = income received from 'rest of the world' by the government
F = total savings in the regional economy
S_k = total household savings
S_i = savings/investment by production sectors
S_g = government savings
S_R = exogenous inflow of capital
M = total value of regional imports
M_k = expenditure on imported goods by household of type k
M_i = value of imported intermediates purchased by sector i
M_g = imports by the government

As in a Leontief system, we assume that all the structural relations (both behavioural and technological) are linear or at least that they can be approximated to linear functions. More specifically, some of these assumptions are stated below.

The model's coefficients were estimated from the SAM entries for 1979-80 for 39 production sectors based on data from Bhalla *et al.* (1990). A description of the various rows of the SAM-based multiplier model is presented in the equations given below.

Households:

$$Y_k = \sum_i w_{ki} X_i + R_k \quad \text{for } k = 1,2,\ldots5.\ i = 1,2,\ldots39$$

[Total income of households of type k] = [Total factor payments received from all production sectors] + [income transfers from abroad to households of type k]

Production Sectors:

$$X_i = \sum_k C_{ik} Y_k + \sum_j a_{ij} X_j + J_i + G_i + E_i$$

[Gross Output of production sector i] = [Sum of all households' demands for output of sector i] + [Intermediate demand by sector j for output of sector i] + [investment demand] + [Government purchases of sector i 's output] + [Exports of sector i]

Government Receipts:

$$G = \sum_k t_k Y_k + \sum_i t_i X_i + R_g$$

[Total tax revenues] = [Taxes paid by all households] + [Taxes paid by all sectors] + [Income received from rest of the world by the government]

Capital Account Receipts:

$$F = \sum_k S_k + \sum_i S_i + S_g + S_R$$

[Total Savings] = [Total household savings] + [Savings/Investment by production sectors]+[Govt savings]+ [exogenous inflow of capital]

Rest of the World Account:

$$M = \sum_k M_k + \sum_i M_i + M_g$$

[Total Imports] = [expenditure on imported goods by all households] + [value of imported intermediates purchased by production sectors] + [imports by the government]

Regional Value-Added:

$$\sum_k \sum_i w_{ki} X_i$$

Regional value-added is the sum of incomes received by all households from all production activities.

In this SAM-based multiplier model of the Punjab regional economy, regional value-added (RVA) has been taken as the objective function[8] to be maximised under the two situations: 'With Project' and 'Without Project'.

Simulations under 'With Project' Situation

As mentioned above, the SAM-based multiplier model described above has been used to compute the values of relevant variables in the 'With Project' situation with their counterparts in the hypothetical case that the project had not been undertaken ('Without Project'). This set of variables comprises all the elements of a SAM for the region in each

8. We have used GAMS (General Algebraic Model System) to obtain optimum solutions.

situation, assuming fixed prices. The optimum solution under 'With Project' situation gives an estimate of the regional value-added alongwith household incomes, sectoral gross outputs or export levels, savings and investments and government expenditure for the year 1979-80.

Simulations under 'Without Project' Situation

The regional value-added (RVA) under 'Without Project' situation has been obtained by using the SAM-based multiplier model by fixing output levels of agricultural commodities and electricity under the hypothetical situation of 1979-80 in the Punjab state in the absence of the Bhakra dam system.

Most of the inter-sectoral interactions in Punjab are driven by output increases in the agricultural sector. Agriculture in turn is driven very strongly by the availability of irrigation water. It is therefore, important to explicitly account for this important driving force of inter-sectoral linkages. To assess therefore, how the non-availability of irrigation water and hydropower in the 'Without Project' situation would have affected the inter-sectoral linkages and the production of different sectors in SAM, it is important to estimate how the agricultural production would have been affected in the event of non-availability of irrigation water and hydropower (a part of which is used for pumping groundwater for irrigating crops).

Irrigation and Agricultural Output under 'Without Project' Scenarios

The major effects of the Bhakra dam system have been in terms of increasing area under irrigation and the output of electricity. Had the dam not been built, it would have affected the availability of irrigation and hydropower. Thus, for the hypothetical case of 'Without Project' scenario, the following assumptions are made with respect to the availability of irrigation and the irrigated area.

Irrigation Under 'Without Project' Scenario in Punjab

The availability of Bhakra waters has helped in several ways in contributing to the increased irrigation availability in Punjab. It has led to changes both in terms of quantity of irrigation water available as well as quality (as measured in terms of reliability of irrigation water at the required time) of irrigation water available. It has helped in extending surface irrigation water availability to hitherto unirrigated areas, in tremendously increasing water availability in the existing canal network and, through seepage of canal waters to the groundwater, in making available huge amounts of groundwater for exploitation through tubewells.[9]

Thus, while it may not be correct to attribute the entire changes on the irrigation scene, presented above, to the building of the Bhakra dam, it is however correct to infer that the irrigation scene in the absence of the project would have been much different than that presented above. In the hypothetical situation of the Bhakra dam not having been built, the likely irrigation scenario in Punjab could be visualised as consisting of the canal network that existed in the pre project period continuing to carry low volumes of irrigation water, though maybe somewhat at a higher level than that in the pre project year, and a reduced level of groundwater availability due to the absence of recharge of groundwater from the canal waters. In the absence of data available on actual amounts of water flowing through the canal system, and to avoid any undue prejudices against the 'without project' scenario, we assume that the volume of water carried in large parts of the canal system would be sufficient to grow all crops including high yielding varieties of crops, except that of Paddy due to its high water requirement. The available water is however, assumed to be sufficient to grow local varieties of paddy due to their low irrigation water requirement. We also assume that of the total canal irrigated area available in 1979-80 in the 'without project' scenario, the water availability in 25 per cent of the cultivated area would be adequate to grow even high yielding varieties of paddy also.

---

9. The role of seeped in waters in improving recharge is often not so well recognised in the case of canal irrigation. A canal system is inherently rather leaky, more so when it is unlined. According to some of the available estimates, the natural groundwater recharge in Punjab could sustain half the existing number of tubewells in Punjab. In other words, the investment in canal works has enhanced groundwater availability in Punjab by a factor of two. Therefore, half the crop output originating from tubewell-irrigated lands in the Punjab State is from groundwater that is of canal origin (Dhawan, 1993). In another study on Punjab, Dhawan (1989) concludes, "These exercises should convince a discerning reader that we are justified to adopt a range estimate of 50 to 70 per cent for the share of canal seepage in total groundwater."

Irrigation from Check Dams

Given the flat topography of the land in most of the command area of the Bhakra dam system, it is assumed that no substantial irrigation could be obtained from investing in small check dams. In any event, the area that could have been potentially irrigated by small dams in the region would have been a very small part of the total area irrigated by large canals from the Bhakra dam system. Hence, we have not made any estimates for the area that could have been irrigated from the small/ check dams under the 'Without Project' situation.

Thus, in the likely scenario of the Bhakra dam system having not been built, we envisage the irrigation scenario in Punjab would have been as follows, depending upon the assumptions made:

Assumption I: It is assumed that the water available in the canal irrigated area is sufficient to grow all crops, except HYV of paddy. However, in 25 per cent of this area, it is assumed that the available water is sufficient to grow HYV of paddy as well. The groundwater irrigated area has been assumed to be 50 per cent of the groundwater irrigated area in the 'With Project' situation and that the entire groundwater irrigated area is suitable for cultivation of all crops, including HYV of paddy.

Assumption II: It is assumed that the water available in the canal irrigated area is sufficient to grow all crops, except HYV of paddy. However, in 25 per cent of this area, it is assumed that the available water is sufficient to grow HYV of paddy as well. The existing groundwater[10] area in the pre-project period will become suitable for HYV of paddy and other crops but there would have been no new additions to area under groundwater irrigation.

The estimated figures suggest that had the Bhakra dam system not been built, the Net Irrigated Area (NIA) would have been between 43

---

10. It is assumed that, in the absence of water releases from the dam, the groundwater recharge would have been lower due to low volume of water available in the canal system. However, the existing wells would have converted into the tubewells and would have drawn more water than they were drawing before the project. But it is assumed that no new area under groundwater would have come. This restrictive assumption has been made to see the impact of low irrigation availability on direct and indirect impacts.

and 54 per cent lower in 1979-80 than that actually prevailing in 1979-80 (Table 6.12).

Table 6.12

Irrigated Area in Punjab under Alternative Assumptions (Thousand Hectares)

	Actual 1950-51	Actual 1979-80	Without Bhakra 1979-80	
			Assumption I	Assumption II
1. Canal Irrigated Area				
(a) Adequate Water for Growing HYV Paddy	–	1496	241	241
(b) Inadequate Water for HYV Paddy	922		681	681
2. Groundwater Irrigated Area	764	2062	1106	764
3. NIA (Net Irrigated Area)	1686	3558	2028	1686
4. GIA (Gross Irrigated Area)	2278	5708	3245	2698

Note: 1. For calculating GIA we have used the same irrigation intensity that prevailed actually in 1979-80 *viz.*, 1.60, though this may somewhat cause an upward bias.

2. While calculating the "without" Bhakra scenario figures we have applied percentage reductions to only Bhakra served area of Punjab.

Agricultural Production under 'Without Project' Scenario

To assess the implication of reduced availability of irrigation water on the agricultural sector, in the hypothetical situation that the Bhakra dam system had not been constructed ('Without Project'), a formal model of the agriculture sector has been developed. In particular, we developed a Linear Programming (LP) model of the agricultural sector, which solves for equilibrium values of the area allocation to different crops for a defined objective function. The model is validated for the year 1979-80. The model simulates the steady state equilibrium as it should have been during that year and may therefore, deviate from the values actually prevailing in the study region. Subsequently the model has been used to analyse the direct effects of the Bhakra project by solving the model for different hypothetical 'Without Project' situations as discussed above.

To simulate the regional agricultural economy in the absence of the Bhakra dam, apart from reduced availability of irrigation water as discussed above, it was also necessary to make some plausible assumptions about some of the underlying variables. It is assumed that identical input-output relations hold under both 'with' and 'without' project situations. It has also been assumed that farmers would have access to the same level of agricultural technology (HYV and mechanisation) and would have used the same input levels such as that of fertiliser and seeds on irrigated farms in the region in the absence of the Bhakra dam as in the presence of the dam. It is also assumed that it was reasonable to foresee the autonomous changes in population, incomes, prices and investments in the absence of the project. Further, we assume that prices of inputs and outputs, subsidies, markets, infrastructure (except that of irrigation), government policies etc. were identical in both situations. Hence, the difference in values of relevant variables derived from the two situations would provide an estimate of the 'pure' impact of the Bhakra dam system. Some of the relevant variables used for this impact evaluation are: increased area under irrigation under rice, wheat, cotton and other crops and the increases in output of these crops as a result thereof.

The production and value of output of some of the major crops, at 1979-80 prices for 'With Project' and 'Without Project' situations, are given in Tables 6.13 and 6.14. It may be noted that Tables 6.13 and 6.14 show the predicted differences between what would have been the situation in the region 'with' and 'without' the Bhakra dam system. In the case of animal husbandry, the value of output under 'Without Project' situation has been estimated to reflect the area available for fodder crops.

The results obtained from solution of the model under different situations suggest that if the Bhakra project had not been undertaken, the production of rice would have been about one half of what was being produced in 1979-80 with Bhakra. Similarly the production of wheat would have been between 54 and 65 per cent of the currently produced quantity of wheat (See Figures 6.8 and 6.9). The total value of these four major crops would have been between 52 and 61 per cent of the value currently obtainable.

Table 6.13

Impact of Alternative Irrigation Availability Scenarios on Production of Some of the Important Crops ('000 Metric Tonnes)

Crop	*LP with 1979-80 Prevailing Conditions*	*LP with Groundwater at 50 per cent of use in 1979-80*	*LP without additional Groundwater use*
Rice	3437 (100)	1717 (50)	1636 (48)
Wheat	8190 (100)	5354 (65)	4438 (54)
Cotton	259 (100)	157 (61)	125 (48)
Sugarcane*	477 (100)	471 (99)	285 (60)

Note: *Production of sugarcane is in terms of *Gur*.

Figures in parentheses are index numbers.

Table 6.14

Value of Output of Major Crops under Alternative Scenarios (Mn Rs. at 1979-80 Prices)

Crop	*LP with 1979-1980 Prevailing Conditions*	*LP with Groundwater at 50 Per cent of use in 1979-80 (Assumption I)*	*LP without additional Groundwater use (Assumption II)*
Rice	6187	3091	2945
Wheat	10647	6960	5769
Cotton	2072	1256	1000
Sugarcane	859	848	513
Total of Four Crops	19765 (100)	12155 (61)	10227 (52)

Electricity Output Under 'Without Project' Scenario

Apart from the reduced availability of irrigation water under the 'without' project scenario, the availability of electricity would have also been affected. We make the following two assumptions about the availability of thermal power in the absence of availability of hydro power from the Bhakra dam system:

Assumption I: No additional thermal power available to replace hydropower available from Bhakra.

Assumption II: Additional thermal power will be available to replace 50 per cent of hydropower available from Bhakra.

Figure 6.8

Production of Different Crops With and Without Bhakra Dam—Thousand Tonnes/yr

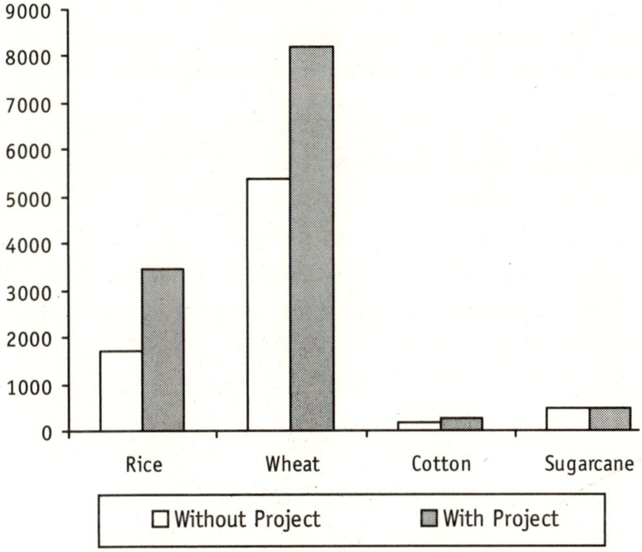

Source: Table 6.13 (Without project refers to Groundwater at 50 per cent availability)

Figure 6.9

Value of Output With and Without Bhakra Dam—Rupees Million/YR

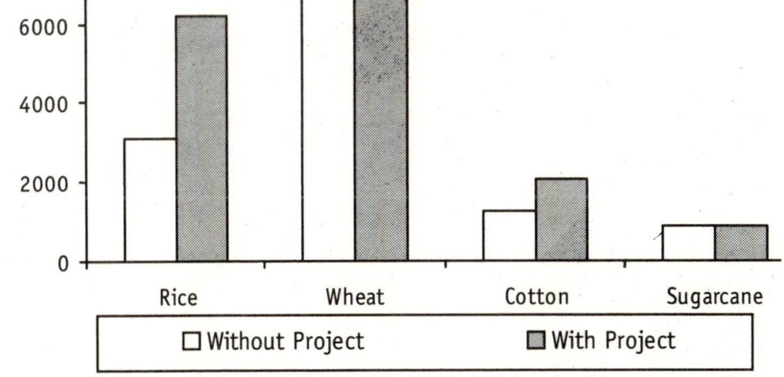

Source: Table 6.14 (Without project refers to Groundwater at 50 per cent availability)

The electricity scenario in the two cases would thus have been as follows:

Assumption I: The installed capacity of electricity in Punjab in 1979-80 was 1536 MW of which 1416 MW was from the Bhakra dam system. In the year 1979-80 total electricity generated in Punjab was 6235 million kWh (mkWh) of which 6033 mkWh was from the Bhakra system.

According to the available data, estimated value of output of electricity, gas and water supply sector in 1979-80 was Indian Rupees[11] (Rs.) 1647 million (Bhalla *et al.*, 1990). Assuming that 90 per cent of this was due to electricity, this gives an estimated value of Rs. 1484 million from electricity for an estimated generation of 6235 million kWh or Rs. 0.24 per kWh. Using this estimate of the unit price of electricity, the estimated value of electricity from Bhakra in Punjab comes to Rs. 1442 million. Thus, in the hypothetical case under which the Bhakra dam system had not been constructed, the value of electricity output in the Punjab would have been Rs. 50 million (Rs. 1484 million minus Rs. 1434 million). Hence, the output value of electricity, gas and water sector in 1979-80 under 'Without Project' situation is estimated at Rs. 213 million.

Assumption II: The value of output from the electricity and gas sectors is taken at 50 per cent of the value of output under 'With Project' situation. Hence, the output value of electricity, gas and water sector in 1979-80 under 'Without Project' situation is estimated at Rs. 823 million.

The above two alternative assumptions have been used to derive simulation results on the impact of non-availability of hydro power from the Bhakra dam system using the SAM model.

Results of Multiplier Analysis

The gross output levels and value-added for Punjab during 1979-80 under different assumptions of irrigation availability and electricity availability in the 'Without Project' situation have been estimated using the SAM model. We present in Table 6.15 results for one such

11. The current (December 2002) conversion rate is Rs 48.4 for a US $. The exchange rate during 1979-80 was approximately Rs 10 to a US dollar. In this report we have used Indian Rupees (Rs) for value figures using 1979-80 prices.

combination of alternative scenarios comprising Assumption I for groundwater and Assumption II for electricity as discussed above. The results are presented for a sectoral aggregation of the 39 sectors. This Table also presents gross output and value-added, respectively, for each aggregated sector as estimated in the SAM for 1979-80 representing the 'With Project' situation. The detailed sector wise gross value of output and value-added for each of the 39 sectors under the 'With Project' situation are presented in Table 6.16.

Under defined combinations of assumptions about the availability of irrigation and electricity under the 'Without Project' situation, the aggregate gross output in the region under the 'With Project' situation is larger by Indian Rupees 20 to 24 billion (30 to 34 per cent) than it would have been had the project not been constructed.

Table 6.15

Estimated Gross Value of Output and Value-Added in Simulation Exercises using the SAM Model for Punjab: 1979-80 (Millions of Indian Rupees)

	Gross Value of Output			*Value Added*		
Sectoral Aggregation	*Without Project*	*With Project*	*Per cent Change*	*Without Project*	*With Project*	*Per cent Change*
Agriculture	15137	22072	46	9870	14398	46
Animal Husbandry	5583	7976	43	3919	5598	43
Agro Processing	3734	5566	49	473	727	54
Manufacturing	17790	21884	23	3866	4767	23
Electricity, Gas and Water	824	1647	100	473	946	100
Construction and Services	27501	30148	9.6	14277	15943	12
Total	70568	89294	26.5	32878	42379	29

Notes: * Figures do not add up to totals due to rounding.

1. In simulations under 'Without Project' scenarios, value of output was fixed for the following sectors: wheat, rice, cotton, animal husbandry and electricity, gas & water (See text).

Assumption: Assuming that under 'without project' situation, groundwater use will be at 50 per cent of the use in 1979-80 (under 'with project' situation) (Assumption I of irrigation availability scenario) and additional thermal power equal to 50 per cent of hydro output (Assumption II of electricity scenario) will be available.

Table 6.16

Value-Added and Gross Output of 39 Production Sectors in the SAM for Punjab (Millions of Indian Rupees) for 1979-80

	Value Added	*Gross Output*	*Value-Added Per Unit of Gross Output*	*Per cent of Regional Value-Added*
Wheat	5421	9735	55.7	12.8
Rice	3574	5189	68.9	8.4
Cotton	1573	2196	71.6	3.7
Sugarcane	650	813	79.9	1.5
Gram and Pulses	365	479	76.2	0.9
Oilseeds	367	454	80.8	0.9
Other Agr. and Forestry	2448	3207	76.3	5.8
Animal Husbandry	5598	7976	70.2	13.2
Dairy Products and Confectionary	155	669	23.1	0.4
Grain Mill Products	157	1874	8.4	0.4
Bakery Products	53	156	34	0.1
Sugar	34	195	17.5	0.1
Other Food Industries	140	512	27.4	0.3
Edible Oil	140	1876	7.5	0.3
Breweries and Beverages	47	285	16.6	0.1
Ginning and Textiles	1251	6932	18.1	3.0
Sawmills and Wooden goods	234	601	38.9	0.6
Paper, Printing and Publishing	51	144	35.3	0.1
Rubber and Leather Products	233	890	26.2	0.6
Basic Chemicals and Fertilisers	271	1055	25.7	0.6
Drugs and Pharma-ceuticals	7	39	17.1	0.0
Other Chemicals	36	233	15.4	0.1
Glass and Mineral Products	561	1488	37.7	1.3
Basic Metal Industries	392	4328	9.1	0.9
Metal Products	280	1377	20.3	0.7
Machinery except Electrical	241	1119	21.5	0.6
Electrical Machinery	112	536	21.0	0.3
Railroad Equipment	28	65	43.3	0.1
Motor Vehicles, Manufacture and Repair	399	1073	37.2	0.9
Bicycles and Parts	412	1313	31.4	1.0

contd...

contd...

	Value Added	*Gross Output*	*Value-Added Per Unit of Gross Output*	*Per cent of Regional Value-Added*
Scientific and Surgical Instruments	25	86	28.8	0.1
Sports and Athletic Goods	102	412	24.7	0.2
Miscellaneous Industries	133	193	68.9	0.3
Electricity, Gas and Water Supply	946	1647	57.4	2.2
Construction	2393	7863	30.4	5.6
Trade, Transport and Storage	8649	13841	62.5	20.4
Other Services	3494	6511	53.7	8.2
Banking and Insurance	877	1123	78.0	2.1
Real Estate and Owner of Dwellings	530	810	65.5	1.3
TOTAL	42379	89294		

As expected, the project had its biggest impact on the output of agricultural commodities, specially the output of wheat, paddy, cotton and oilseeds. The output of agricultural commodities is larger by Rs. 7 to 9 billion (46 to 66 per cent) under the 'With Project' situation than it would have been had the project not been undertaken. The output of electricity is estimated (for 1979-80) to be higher by 6033 million kWh or Rupees 1442 million.

Table 6.17 presents the results of the estimates of value-added multipliers based on the methodology and assumptions described above. Two estimates of multipliers have been reported based on the assumptions regarding the impact of seepage from canals on the availability of groundwater irrigation and the availability of electricity. These are:

Scenario I: Assuming that under the 'without project' situation, groundwater use will be at 50 per cent of the use in 1979-80 (Assumption I of irrigation under 'Without Project' situation) and additional thermal power equal to 50 per cent of hydro output will be available (Assumption II of electricity under 'Without Project' situation).

Scenario II: Assuming that under 'without project' situation, groundwater use will be at 50 per cent of the use in 1979-80 (Assumption I of irrigation under 'Without Project' situation) and no

Table 6.17

Estimated Values for Multiplier Effects of the Bhakra Dam Project, India

Definition of Multiplier	$\frac{\text{Direct Effects} + \text{Indirect Effects}}{\text{(Direct Effects)}}$
Value (Millions of Indian Rupees (Rs.) Scenario I	$\frac{\text{Regional Value Added (With Project)} - \text{Regional Value Added ('Without Project')}}{\text{Value Added of Ag.\& Elec. ('With Project')} - \text{Value Added of Ag.\& Elec. ('Without Project')}}$ $\frac{42379 - 32878}{15343 - 10343} = 1.90$
Value (Millions of Indian Rupees (Rs.) Scenario II	$\frac{\text{Regional Value Added (With Project)} - \text{Regional Value Added ('Without Project')}}{\text{Value Added of Ag. \& Elec. ('With Project')} - \text{Value Added of Ag. \& Elec. ('Without Project')}}$ $\frac{42379 - 30729}{15343 - 8807} = 1.78$

Notes: Scenario I: Assuming that under without project situation, groundwater use will be at 50 per cent of the use in 1979-80 (under with project situation) and additional thermal power equal to 50 percent of hydro output will be available.

Scenario II: Assuming that under without project situation, groundwater use will be at 50 per cent of the use in 1979-80 (under with project situation) and No additional power available from thermal sources

Source: See text and tables

additional thermal power is available in the absence of hydropower from Bhakra system (Assumption I of electricity under 'Without Project' situation).

Under Scenario I, the results show that the regional value-added under 'With Project' scenario at Rs. 42.38 billion is larger than the regional value-added under 'Without Project' scenario by Rs. 9.5 billion. Compared with this, the value-added from agriculture and electricity sectors at Rs. 15.34 billion is larger than the corresponding value-added under the 'Without Project' scenario by Rs. 5 billion. Thus, in the aggregate, the project induced an increase of Rs. 9.5 billion in regional value-added. Of this, Rs. 5 billion is due to an increase in the outputs of agricultural commodities and the additional generation of hydroelectricity from the Bhakra dam.

This gives a multiplier value[12] of 1.90 i.e. Rs. 9.5 billion/Rs. 5 billion. Thus, for every rupee (100 paise) of additional value-added

12. Here we are using the multiplier value for the project as a whole as defined in Bell *et al.* (1982). An alternative estimate of multiplier, as used in regional science literature, would be higher at 2.78 when total effects *vis-à-vis* direct effects are considered under the 'With Project' situation.

directly by the project in agricultural and electricity sectors, another Re 0.90 (90 paise) were generated in the form of downstream or indirect effects.

Under Scenario II, the results show that when electricity availability is lower than in estimate I (under the scenario of 'Without Project'), the regional value-added under 'With Project' scenario at Rs. 42.38 billion is larger than the regional value-added under 'Without Project' scenario by Rs. 11.65 billion. Compared with this, the value-added from agriculture and electricity sectors at Rs. 15.34 billion is larger than the corresponding value-added under 'Without Project' scenario by Rs. 6.53 billion. This gives a multiplier value of 1.78 i.e. Rs. 11.65 billion/ Rs. 6.53 billion. Thus, for every rupee of additional value-added directly by the project in agricultural and electricity sectors, another Re 0.78 were generated in the form of downstream or indirect effects.

Thus, the simulation results based on a SAM model show that the Bhakra dam project generated significant indirect or downstream effects in Punjab state. The multiplier values range between 1.78 to 1.90 depending on the assumptions regarding the impact of canals on groundwater availability. For every rupee (100 paise) generated directly, another 78 to 90 paise were generated in the region as downstream or indirect effects. These multipliers include the effects of inter-industry linkages as well as consumption-induced effects.

Multiplier Effects Outside the Region

In addition to the direct and indirect effects realised in the directly impacted region falling in the Bhakra system command, the dam has directly and indirectly also impacted regions and people in area located far-off from the project region. The increased economic activity in the directly impacted region has provided employment opportunities to thousands of migrant labours from some of the poorest regions of the country such as from the states of Bihar and Uttar Pradesh. The remittances from these migrant labours back to their families in their native villages in turn has impacted the economy of these villages. The huge foodgrain surplus production from the project region is procured by Government agencies for distribution at relatively low prices through 'fair price shops' to urban consumers all over the country. These procured foodgrains have also been used for giving wages 'in kind' to labours employed under food-for-work programmes etc. carried out in different parts of the country.

We briefly discuss some of these impacts.

Multiplier Effects of Remittances by Migratory Labourers in Villages Outside Punjab

The shifts in the cropping pattern and attendant changes in crop production brought about by the availability of irrigation water in the region created huge employment opportunities in agriculture and related sectors. The employment opportunities available on a continuing basis for hired labour every year lures thousands of labour from far-off poor regions of the country—Bihar, Uttar Pradesh etc, where wage rates are low and unemployment is very high, to migrate to this region in search of employment and better wages. While some of these labours have settled down permanently in the region itself, others migrate every year. In a study carried out recently by the Punjab Agricultural University it has been estimated that during the lean period of agricultural operations the number of migrant labour employed in Punjab was 387 thousand and this number increased to about 774 thousand in the peak period (Sidhu *et al.*, 1997). About 93 per cent of the migrants belonged to the poor states of Bihar and Uttar Pradesh while about 5 per cent also came from Nepal. It has further been estimated that the number of migrant labour have increased by about 35.31 per cent in 1995-1996 as compared to 1983-84. In terms of the proportion of migrant labour to local labour the proportion of migrant to local labour increased from 7.60 per cent in 1978-79 to 11 per cent in 1995-96. In other words, Punjab agriculture meets almost one-tenth of its total labour requirement by migrant workers.

The wage rates, which these migrant labourers get in their native villages, are much lower than what they get in Punjab. Thus, for example, about 46 per cent of migrant labour were getting less than Rs. 300 per month for being employed on a permanent basis in their native villages while in the Punjab they were getting almost 200 per cent higher wages.

It has been estimated that total earnings of the entire migrant labour force in crop production in Punjab during 1995-96 was Rs. 5344 million (US$ 114 million) out of which they remitted Rs. 3548 million (US$ 75 million or 66 per cent) back to their native places while the remaining Rs. 1796 (US$ 39 million) were spent by them in Punjab itself. About 18 per cent of the migrant labour utilised their savings for creation of assets in their native places.

The remittance of such huge amounts of money to relatively far-flung poor regions from where this labour often comes has helped improve the living conditions of people living there and has created its own further downstream effects. About 96 per cent of the migrant labours reported that they utilised these savings for purchasing consumer durables such as televisions, radios, bicycles, sewings machines etc. About 18 per cent also utilised their savings from Punjab on the repair and construction of their houses in their native village. A small percentage of labourers also used these savings to either lease land for increasing their operational holdings or for buying a new piece of agricultural land. In addition, part of the savings were of course used for improving daily consumption.

Apart from earning cash, these migrant labourers have acquired new skills in crop cultivation and operation of farm machinery during their stay in the Punjab. This has offered a great scope for transformation of the agricultural economies of their native villages but these labourers have not been able to use their acquired skills in their native villages due to a number of factors such as the non-availability of irrigation, inadequate credit etc. Nevertheless migration has helped them improve their skills

Multiplier Effects of Surplus Foodgrains used in 'Food for Work' Programmes

During the year 2001-02 the States of Punjab and Haryana contributed 17 million tonnes of wheat out of a total of 20.63 million tonnes procured by the State foodgrain procurement agencies from all over the country. Similarly during the year 2000-01 out of a total procurement of 19.10 million tonnes of rice from the entire country, Punjab contributed 6.93 million tonnes.

The Bhakra irrigated region provides the bulk of the foodgrains required by the Central Government for running the food distribution system in the country. While part of the procured foodgrains are utilised for running the Public Distribution System (PDS) of the country, a part of the procured foodgrains are provided to the States during emergencies created by natural calamities such as floods, droughts etc. These allocations in part are used by the States to create gainful employment opportunities under various employment generation programmes such as for creation of infrastructure etc. in which part of the wage payments are made using these foodgrains. For example during 1997 the central Government allocated 1303 thousand tonnes of rice and 705 thousand tonnes of wheat for various employment creation programmes in the States.

Multiplier Effects of Water Supply to Rural and Urban Areas

As mentioned earlier, the Bhakra system has provided adequate drinking water supply to many towns and cities in the states of Punjab, Haryana and Rajasthan. Even Delhi gets a part of its water supply from the Bhakra system.[13] In the absence of this water supply, the growth of industry and services in these towns and cities would have been adversely affected. However, due to the non-availability of any studies on the subject, it has not been possible to make any estimates of direct or indirect (and multiplier benefits) from the availability of water supply that is critical for generating urban incomes and employment.

Income Distribution Impacts of the Bhakra Dam System

The availability of irrigation water, and the consequent growth in rural and urban areas that followed, has helped in a very significant way in bringing down poverty in the two states of Punjab and Haryana. As already mentioned the benefits of this economic development in the region have been shared both by rural as well as urban areas. The income distribution effects of the Bhakra dam have been analysed using

13. Total water supply to Delhi from Bhakra in 1999 was 2803 cumec days out of a total of 428182 cumec days. However, the water supply from Bhakra meets a substantial part of the industrial demand for water supply.

the SAM-based multiplier model for the Punjab economy described above.

The income distribution effects of the Bhakra dam have been analysed in three ways:

(i) By comparing the shares of agricultural labour and other households in aggregate incomes under 'With' and 'Without Project' in the Punjab state.

(ii) By comparing the differences in aggregate income levels of various household categories under two scenarios of 'Without Project' and 'With Project'.

(iii) By assessing direct and indirect components of income differences under 'With' and 'Without Project'.

Shares of Agricultural Labour and Other Households in Aggregate Incomes under 'With' and 'Without Project'

The total population of 12.7 million in the Punjab has been divided into urban households (0.7 million) and four categories of rural households, namely, households self-employed in agriculture[14], agricultural labour households, non-agricultural labour households and other households. Of the total 12 million rural population, 46 per cent or 5.5 million were self-employed in agriculture while 23 per cent or 2.7 million were agricultural labour households. Non-agriculture labour households constituted about 6 per cent and other rural households constituted about 25 per cent of total rural households.

The SAM-based multiplier models have been used to assess the income effects of additional irrigation and hydropower provided by the Bhakra dam. This has been done for each of the above defined household categories.

Figures 6.10 and 6.11 depict results of the shares of different category of households in the aggregate rural income under 'with' and

14. Those households which earn 50 per cent or more of their total income from self-employment in agricultural occupation during the last 365 days are treated as self-employed in agriculture. Those households which earned 50 per cent or more of their total income during the last 365 days from wage paid manual labour in agriculture are treated as agricultural labour households.

Figure 6.10

Income Shares: Without Project

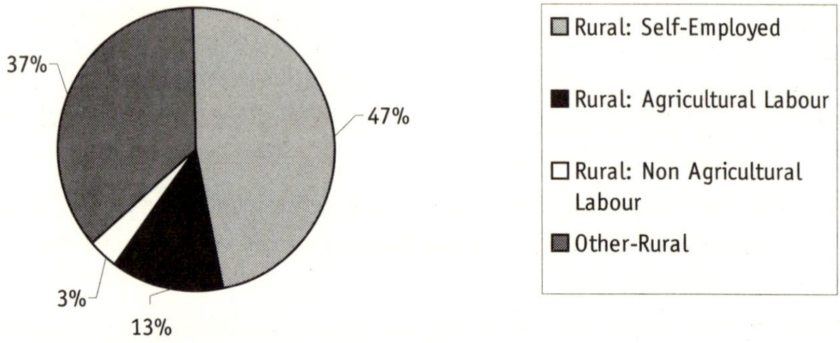

Source: Author's calculations and based on model's results.

Figure 6.11

Income Shares: With Project

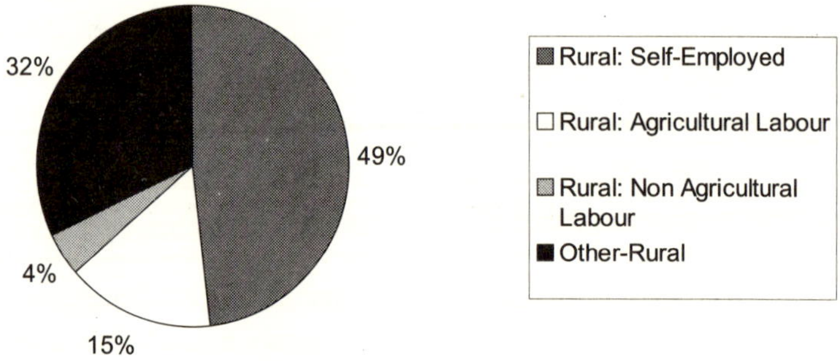

Source: Author's calculations and based on model's results.

'without project' situations. For example, for the self-employed rural households, the share of their income in aggregate rural income levels is 46.8 per cent under the 'without project' situations compared with 48 per cent under the 'with project situation'. The share of agricultural labour households at 15.4 per cent of the total under 'with project' situation is higher than 12.9 per cent under 'without project' situation. However, when a comparison is made between shares of aggregate rural and urban incomes in the total, the results show that the share of rural households is higher (61.5 per cent) under the 'with project' situation when compared with that (57.4 per cent) under the 'without project' situation. The share of urban households is 42.6 per cent and 38.5 per cent in the aggregate income under the 'without project' and 'with project' situations respectively.

Differences in Aggregate Income Levels of Various Household Categories under 'Without Project' and 'With Project'

The SAM-based multiplier models have been used to assess differences in household incomes under the 'with' and 'without project' situations.

Table 6.18 and Figures 6.12 and 6.13 show that all households have higher income levels under 'With Project' situation than under 'Without

Table 6.18

Differences in Incomes of Agricultural Labour and Other Rural Households: With and Without Bhakra Dam (Rupees Million)

Category Of Households	*With Project*	*Without Project*	*Difference*	*Percent Difference*
Self employed Rural Households	12505	8825	3680	41.7
Agriculture labor	4005	2425	1580	65.2
Rural Non Agriculture	1125	627	498	79.5
Rural-Others	8413	6988	1425	20.4
Urban Households	16331	14014	2317	16.5
TOTAL	42379	32878	9501	28.9

Source: Simulations using the SAM model for the Punjab 1979-80 described in the text (See text).

Figure 6.12

Income of Different Types of Households With and Without Project (Rs. Million)

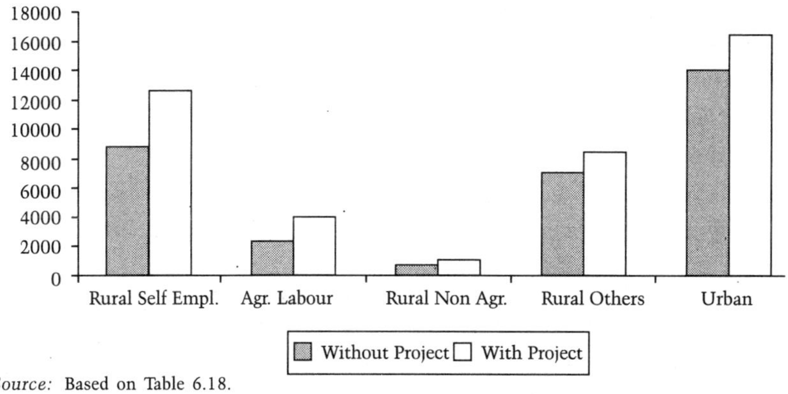

Source: Based on Table 6.18.

Figure 6.13

Percentage Change in Income of Different Types of Households With and Without Bhakra Dam

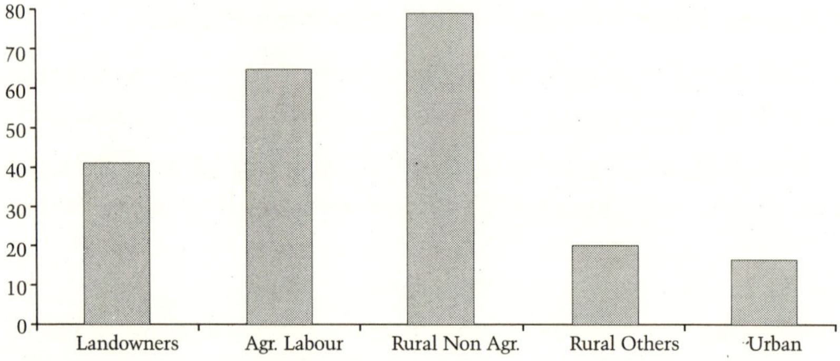

Source: Based on Table 6.18.

Project' situation. For example, for self-employed rural households, the difference in income levels (under 'with' and 'without project' situations) is estimated to be Rs. 3680 million. Their income under 'with project' situation is 42 per cent higher over the income level under 'without project' situation (Rs. 8825 million).

In the case of agricultural labour households, the difference in income is Rs. 1580 million, which shows that their income under 'with project' situation is 65 per cent higher than the income level of Rs. 2425 million under 'without project' situation.

However, in the case of urban households, the difference in income levels under the two situations is relatively low, only 16.5 per cent. The level of income of urban households under 'without project' situation is Rs. 14014 million compared with Rs. 16331 million under the 'with project' situation. The results show that the gain for rural households from the dam is relatively higher than the average difference of 29 per cent between aggregate incomes under 'with' and 'without project' situations. Further, the investment in the Bhakra dam has provided income gains to agricultural labour households that are higher than those for the average households.

Direct and Indirect Components of Income Differences under 'With' and 'Without Project'

The SAM-based multiplier models have also been used to estimate direct and indirect income effects of additional irrigation and hydropower provided by the Bhakra dam. This has been done for each of the five defined household categories. The SAM coefficients for each household and each production sector have been used to estimate the value-added received by each household category from each sector. These SAM coefficients have been used to estimate the component of a household's income that is due to value-added from sectors affected directly by the outputs of the dam, namely irrigation and hydropower. Similarly, SAM coefficients for remaining production sectors have been used to estimate the component of income that is due to value-added from sectors affected indirectly by the outputs of the dam. This type of analysis is based on the usual assumptions for this type of fixed-price SAM-based multiplier models, namely that the value-added per unit of output remains the same for each sector and the adjustments take place through changes in quantity. The procedure used to estimate the share of direct and indirect components in household incomes is as follows. First, the SAM coefficients have been used to estimate values of a new SAM for the 'Without Project' situation by fixing the outputs of sectors directly affected by the dam (agricultural sectors and electricity). Then the difference between the values of outputs and household incomes in the two SAMs is used to estimate the differences in incomes under the 'with project' and 'without project' situations. The difference in household incomes is further divided into two components—one from direct sectors and the other from indirect sectors. To calculate the component from direct sectors we look at the row of each household in the matrix of income differences and sum up the incomes the household received from sectors directly affected by the project namely—wheat, rice, sugarcane, cotton, gram, other agriculture and electricity. To calculate the component from the indirect sectors we added the incomes received from the remaining production sectors out of the total of 39 production sectors used in the SAM.

The resulting estimates show, as predicted, that a major part of the income of rural households comes from the output of sectors directly

Figure 6.14

Income Gains to Households from Sectors Affected Directly and Indirectly by the Dam

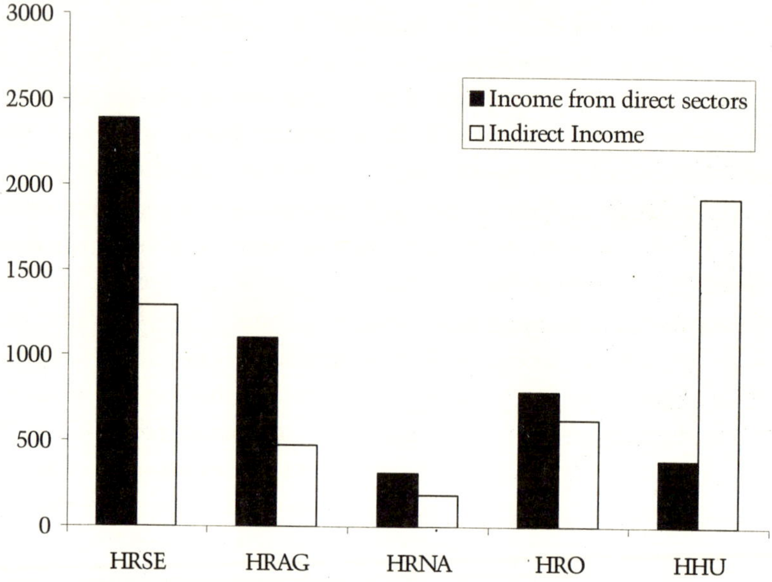

Source: Author's Calculations

affected by the dam. Figure 6.14 presents the results of the disaggregation of the differences in household incomes under the 'with' and 'without project' situations into sectors affected directly by the dam and those indirectly affected. For example, for the self-employed rural households, the difference in income levels (under 'with' and 'without project' situations) is estimated to be Rs. 3680 million. About two-thirds of this difference (or Rs. 2391 million) is due to higher output and value-added in sectors that are directly affected by the dam, namely agricultural crops and electric power. The rest, about one-third, of the difference in income (or Rs. 1288 million) is attributable to changes in value of output and value-added of sectors that are affected indirectly (through inter-industry linkages and consumption-induced effects). In the case of agricultural labour households, as much as 70 per cent of the difference in income is due to changes in output/value-added of sectors that are directly affected by the dam. However, in the case of urban households, the situation is completely reversed. As much as 83 per cent (or Rs. 1925 million) of the difference (under 'with' and 'without project' situations) is accounted for by changes in outputs of sectors that are indirectly affected by the dam, namely agro-processing, manufacturing of textiles etc.

Social and Environmental Effects

Displacement of People and their Rehabilitation

Construction of big dams have almost always led to the submergence of large areas and displacement of people and the studied Bhakra dam is no exception. The figures about the number of actual people displaced by different dams vary very substantially according to the source of information.

The Bhakra reservoir submerged 17,800 hectares of land, affecting 371 villages with a population of about 36,000 people or about 7,200 families. Bilaspur town with a population of 4000 was also submerged. Of the 7,206 affected families, 5,027 were given cash compensation under the Land Acquisition Act and resettled elsewhere on their own. A Bhakra Rehabilitation Committee was appointed to examine the future of the remaining 2,179 families. Since these farmers expressed a desire for resettlement in the Bhakra command, the committee located suitable lands in compact blocks in Hissar and Sirsa districts in Punjab (now Haryana) and 13,200 acres were accordingly acquired in 30 villages. As the amount due in compensation exceeded the value of the land acquired, the displaced persons were compensated partly in land and partly in cash.

Landless tenants were declared eligible to be given an area equivalent to their submerged tenancy, subject to a maximum of 5 acres. Other artisans and labourers from the affected villages were offered a half acre homestead plot free of cost. Those choosing to remain in Himachal Pradesh were granted free fishing licenses in Gobindsagar for three years (Verghese, 1994)

The Pong Dam displaced 16,000 families (about one hundred thousand people). An attempt was made to rehabilitate about half of them in the faraway deserts of Rajasthan in the command area of the project. Each family was given 16 acres of land (the highest so far under any rehabilitation scheme in the country). In spite of this, unable to adjust to the new climatic conditions, water, people and language, most of the displaced people sold their lands and returned to their native place.

In a recent press statement, quoting government sources, it was reported that there are still 784 oustees of Bhakra dam who still remain to be rehabilitated.

Health Impacts

The availability of dependable source of drinking water for millions both in the project region as also in the neighbourhood is likely to translate into better health of people. With the Indira Gandhi canal, the percentage of the reported relative incidence of waterborne diseases to all diseases dropped considerably over a 12-year period in areas that received continuous irrigation facilities (Goel, 2000). In addition, the improvement in economic conditions of a majority of population within the irrigation command helps investment in improved sanitation and also makes the people more health conscious resulting in the reduced incidence of diseases.

Environmental Impacts

The construction of dams, apart from causing economic impacts, also causes a number of social and environmental impacts, some of which are beneficial for the environment while others have negative impacts.

Sedimentation of Reservoirs

The sedimentation of reservoirs as a result of soil erosion in the catchment areas of these reservoirs has been much faster than that assumed at the time of planning of these reservoirs. The Ravi, Beas and Sutlej have substantial upper catchments, the Pong catchment extends over 12,560 square kms. The Sutlej catchment of 56980 square kms is much larger. Approximately 27,000 acre-feet of silt flushes down the Sutlej annually, as measured near the Bhakra. The Bhakra dam was originally based on an assumed siltation rate of 4.29 ha-metre per 100 square kms but this had increased to 6.22 by 1979. Similarly in the Beas reservoir, the sedimentation rate during 1981 was 23.58 as against the original rate of 4.29 assumed during 1974 (NLUPB, 1988)

The faster than assumed siltation of reservoirs has resulted in a serious reduction in the carrying capacity of these structures and has accordingly reduced the efficacy of these reservoirs. The Gobindsagar Lake started filling in 1959 and up to the end of 1988 had lost 21.99 per cent of dead and 6.42 per cent of live storage with an overall capacity loss of 10.26 per cent. The total silt deposit in Gobindsagar in 28 years up to 1987 was 0.935 MAF, the average up to 1986 working out to 25,300 acre feet per annum as against the design figure of 27,250 acre feet.

The Sixth sedimentation report of the Pong Dam, 1988-89, indicated an average sedimentation rate of 26,500 acre feet per annum as against the designed figure of 20500 acre feet. The total silt deposit in the reservoir between 1974 and 1989 worked out to 0.3975 MAF or 5.72 per cent of the gross storage capacity. The small Pandoh reservoir, however, had accumulated 16,886 acre feet of silt as against a gross storage capacity of 33,240 acre feet as of October 1989 (BBMB, 1990)

Waterlogging and Soil Salinity

Large dams have diverse and complex ecological impacts and they often generate environmental costs for those very groups who are supposed to be the beneficiaries. Water logging and salinisation are twin problems caused by the wasteful use of water.

Waterlogging is caused by the interaction of a large number of factors such as irrigation intensity, soil characteristics, drainage, seepage from reservoirs, distributaries and field channels and becomes almost inevitable in areas with undulating topography and water retentive soils.

According to an assess-ment of the Ministry of Water Resources, the extent of waterlogged and soil affected areas in the irrigation command area of Punjab is 200 thousand hectares of waterlogged area and 490 thousand hectares of salt affected area. Similarly in Haryana the corresponding figures were 249 and 197 thousand hectares while in Rajasthan the figures are 180 and 70 thousand respectively (Government of India, 1991)

Groundwater Pollution

The availability of Bhakra water supplemented by groundwater led to adoption of intensive cultivation practices based on intensive use of agro

Table 6.19

Nitrate Content of Groundwater in North West India: 1995

State/Nitrate (mg/l)	*Content*	*Number of Samples*	*Average Nitrate Level*
Punjab			
0-45		328 (68.9)	14.50
46-100		71 (15.1)	66.49
>100		71 (15.1)	238.97
Total		470 (100)	56.26
Haryana			
0-45		222 (63.1)	15.41
46-100		62 (17.6)	68.58
>100		68 (19.3)	283.65
Total		352 (100)	85.59
N-W India			
0-45		550 (66.9)	14.84
46-100		133 (16.2)	67.46
>100		139 (16.9)	280.82
Total		8229 (100)	64.97

Note: Figures in parentheses denote percentages

Source: Singh and Bal (1987).

chemicals in the region. The indiscriminate and unsupervised use of agro chemicals (fertilisers, pesticides and insecticides etc) have been one of the important sources of non point pollution. In addition to loss through run-off, these chemicals leach in to the groundwater aquifer and have the danger of polluting the groundwater.

Table 6.19 shows that the desired level of nitrate in large parts of Punjab and Haryana groundwater has already reached a level above that permissible in drinking water. The results show that 33 per cent of the samples in North-West India had nitrate content well above the permissible limit of 45 mg/l. In fact, in about 17 per cent of the samples, the nitrate content was more than 100 mg/l.

Making Deserts Green

Large dams can also become instruments for improving the environment, as has been the case in Western Rajasthan, which has been transformed in to a green area because of the Indira Gandhi Canal, which draws water from Bhakra Nangal dam. The project has not only allowed the farmers to grow crops in deserts but also has also checked the spread of the Thar desert in adjoining areas of Punjab and Haryana. Besides providing irrigation benefits for over 800 thousand hectares of desert land, which now cultivates crops like wheat, cotton, mustard etc, the most needed water (11 per cent of the project water) is reserved for drinking, domestic, energy and industrial uses in areas around Bikaner, Jodhpur, Jaisalmer towns and 3500 villages. The value of agricultural produce from this vast new areas brought to cultivation is nearly Rs. 12,000 million which is likely to double up as and when the project is complete (Gopalakrishnan, 2000).

Restoration of Loss due to Forest Submergence

The loss in biomass, due to submergence of forest land by reservoirs created for the storage of water, has been more than compensated by the greening of thousands of hectares of land and increased biomass production resulting from increased agricultural production in the project region.

Mitigation of Green House Gas Emission

If the environmentally clean hydro power of 14110 million kWh generated during 1998-99 were to be produced from the next best alternative of coal based thermal power plant, the plants would have emitted, 106 million tonnes of carbon dioxide per year in addition to emitting several other green house gases (taking the carbon dioxide emission from coal based thermal power plants to 330 kg/GJ). The creation of hydropower potential has thus enabled mitigation of GHGs to the tune of 106 million tonnes of carbon dioxide. However, there are other effects of greenhouse gas emissions of reservoirs and rice farming that have to be taken into account to get the net effect of the dam on GHG emissions.

Increased Emission of Methane

The cultivation of wetland paddy is one of the important sources of methane emission. Methane emission from paddy cultivation accounts for about 25 per cent of the total methane emissions from all sources.

The large-scale diversion of cropping pattern in favour of cultivation of rice has thus led to the increased emission of methane from rice fields.

Seismic Impacts

Doubts are often raised about the safety of large dams due to seismic forces. It has been argued that impounding water in reservoirs induces seismicity. The Bhakra dam which has been constructed in the Himalayan region for more than 40 years now has not led to any increase in seismic activity after construction.

Conclusions

The simulation results based on a SAM-based multiplier model show that the Bhakra dam project generated significant indirect or downstream effects in the Punjab state. The multiplier values ranged between 1.78 to 1.90 depending on the assumptions regarding the impact of canals on groundwater availability and availability of power. For every rupee (100 paise) generated directly, another 78 to 90 paise were generated in the region as downstream or indirect effects. These multipliers include the effects of inter-industry linkages as well as consumption-induced effects.

The multiplier effects will be much higher if indirect and induced effects of remittances sent by agricultural workers from Punjab and contributions of the Bhakra dam towards the 'Food for Work' programme are also included in the analysis.

The results on income distribution show that the gains to agricultural labour households from the dam are higher than gains to other rural households and to the urban households. For example, for the agricultural labour households, the income level under 'With Project' situations is estimated to be 65 per cent higher than the income level under the 'Without Project' situation. The corresponding figures for self-employed rural households and urban households are 42 per cent and 16 per cent, respectively.

References

BBMB (2000). *Annual Administration Report 1998-99.* Chandigarh: Bhakra Beas Management Board.

———. (1990). *Annual Administration Report 1989-90.* Chandigarh: Bhakra Beas Management Board. March.

———. (n.d.). *Consultancy Services.* Chandigarh: Bhakra Beas Management Board.

———. (n.d.). *Corporate Brochure.* Chandigarh: Bhakra Beas Management Board.

Bell, C., P. Hazell, and R. Slade (1982). *The Project Evaluation in Regional Perspective A Study of an Irrigation Project in Northern Malaysia.*

Bell, C. and S. Devarajan (1980). "Semi-Input-Output and Shadow Prices: A Critical Note", *Oxford Bulletin of Economics and Statistics*. August.

Bhalla, G.S., G.K. Chadha, S.P. Kashyap and R.K. Sharma (1990). *Agricultural Growth and Structural Changes in the Punjab Economy: An Input-Output Analysis.* Washington D.C.: IFPRI

Bhatia, R., M. Scatasta, and R. Cestti (2003). *Study on the Multiplier Effects of Dams: Methodology Issues and Preliminary Results.* Presented at the Third World Water Forum. Kyoto, Japan. March 16-23.

Chopra, Kanchan (1972). *Dualism and Investment Patterns: An Analysis of Regional Contrasts*. New Delhi: Tata McGraw Hill.

Dhawan, B.D. (1989). *Studies in Irrigation and Water Management.* Delhi: Commonwealth Publishers.

———. (1993). *Trends and New Tendencies in Indian Irrigated Agriculture*. New Delhi: Commonwealth Publishers

Goel, R.S. (2000). "Big vs. Small Dam Controversy: A Critical Overview of Socio-Economic and Environmental Concerns in Indian Context", Paper presented at the International Conference on *Sustainable Development of Water Resources*. New Delhi. November 27-30.

Gopalakrishnan, M. (2000). "The Role of Large Dams in India", Paper presented at the *International Conference on Sustainable Development of Water Resources*. New Delhi. November 27-30.

Government of Haryana (various years). *Statistical Abstract of Haryana*. Chandigarh: Office of the Economic and Statistical Adviser.

Government of India (2002). *Economic Survey 2000-01*. New Delhi: Planning Commission.

———. (1991). *Report of the Working Group on Water Logging, Soil Salinity and Alkalinity.* New Delhi: Ministry of Water Resources.

Government of Punjab (various years). *Statistical Abstract of Punjab*. Chandigarh: Office of the Economic and Statistical Adviser.

Government of Rajasthan (various years). *Statistical Abstract of Rajasthan*. Jaipur: Office of the Economic and Statistical Adviser.

Hazell, Peter B.R. and C. Ramasamy, (1991). *The Green Revolution Reconsidered: The Impact of High-Yielding Rice Varieties in South India.* Baltimore, MD: The Johns Hopkins University Press.

Minhas, B.S., K.S. Parikh, T.N. Srinivasan, S.A. Marglin and T.E. Weisskopf (1972). *Scheduling the Operations of the Bhakra System.* Calcutta: Statistical Publishing Society.

NLUPB (1988). *Report of the Committee of Experts on the Draft Outline of the National Land Use Policy.* New Delhi: National Land Use Planning Board.

Raj, K.N. (1960). *Some Economic Aspects of the Bhakra Nangal Project.* Bombay and London: Asia Publishing House.

Rangachari, R. (2000). "Contribution of Large Dams in India", Paper presented at the *International Conference on Sustainable Development of Water Resources.* New Delhi. November 27-30.

Rao, P.S. and S.Ramaseshan (1985). *Study of Bhakra Reservoir Operation.* Vol. 8, Part 2. Sadhna: Indian Academy of Sciences.

———. (1985a). *Integrated Operation of Beas-Sutlej System.* Vol 8, Part 2. Sadhna: Indian Academy of Sciences.

Sidhu, M.S., P.S. Rangi and Karam Singh (1997). *A Study on Migrant Agricultural Labour in Punjab.* Ludhiana: Punjab Agricultural University, Department of Economics and Statistics.

Singh, B. and H.S. Bal (1987). "Indiscriminate Fertilizer Use *vis-s-vis* Groundwater Pollution in Central Punjab", *Indian Journal of Agricultural Economics* 42(3): 404-09.

Verghese, B.G. (1994). *Winning the Future: From Bhakra to Narmada, Tehri and Rajasthan Canal.* Delhi: Konark Publishers.

7 Check Dams in the Bunga Village, India

R.P.S. Malik and Meera Bhatia

Introduction and Overview

The Shivalik hills, running between the productive and well-irrigated plains of North-West India and the high Himalayan ranges, are composed of heavily erosive, loosely consolidated sandstones and gravel. People living in the lower Shivalik range of the Himalayas, were very poor and had been witnessing severe environmental degradation of the region and experiencing the effects of this degradation on their living conditions. The region was characterised by low agricultural productivity, compounded by a lack of any irrigation facility. A small check dam built in one of the villages in the region, Sukhomajri, demonstrated the change that harvesting of rainwater can bring about in arresting environmental degradation and in bringing about an improvement in the living conditions of the poor people in the region.

The success of the Sukhomajri water harvesting structure led to the construction of a number of similar other structures in the surrounding villages of the Himalayan Shivalik. During the two-decade period between 1976-1996, 102 water harvesting structures were constructed in 60 villages of the Shivalik. This paper describes some of the direct and indirect impacts derived from the water harvesting structures constructed in one of these villages, Village Bunga, and attempts an estimation of the multiplier effects of these dams. While Sukhomajri was the first watershed to be built in the area, Bunga was the third to be constructed.

Check Dams in Village Bunga

Two check dams (Figure 7.1) were built in Village Bunga—the first in 1984 and the second in 1986. The second dam however, could not deliver water due to some bureaucratic hurdles in laying of pipelines for distribution of the water. The second dam actually started supplying water in the year 1996. The two structures combined together are presently supplying irrigation water to 276 ha of village cultivated land. The main features of the two check dams of Village Bunga are presented in Table 7.1.

Table 7.1

Main Features of the Check Dams in the Bunga Village

	Bunga	
	Dam 1	*Dam 2*
Year of Construction/Availability of Water	1984	1996*
Catchment Area (ha)	155	101
Dam Height (m)	16	15
Storage Capacity (ha meters)	180	215
Pondage (ha)	11.1	12.4
Command Area: Proposed (Ha)	243	268
Command Area in 2002 (Ha)	156	120
Total Cost (Thousand Rs. in 1984 prices)**	3951	4431

Notes: *–Although the dam was completed in 1986 the distribution pipelines could not be laid and the dam actually started giving water from 1996.

**–In 1984 1 US$ = Rs. 10.5 Appx.

Source: Arya and Samra (2001).

Major Outputs of the Dams

The major outputs of both the check dams have been making water available for irrigation. However, since the requirements of water for a fully irrigated agriculture in the village are much more than the water, these small dams can store and supply, these dams have been able to fulfill only a part of the irrigation water requirements of the village. To make more efficient and equitable use of the available water, the water supply has been restricted both in terms of the amount of water that an individual household can get as also the season in which the water supply can be made available. The magnitude of shifts that have thus, occurred in the agricultural production scenario have been less than the potential and have been contingent upon the availability of water.

Figure 7.1

Location of Sukhomajri and Bunga Check Dams in Shivalik Hill Region in Haryana, India

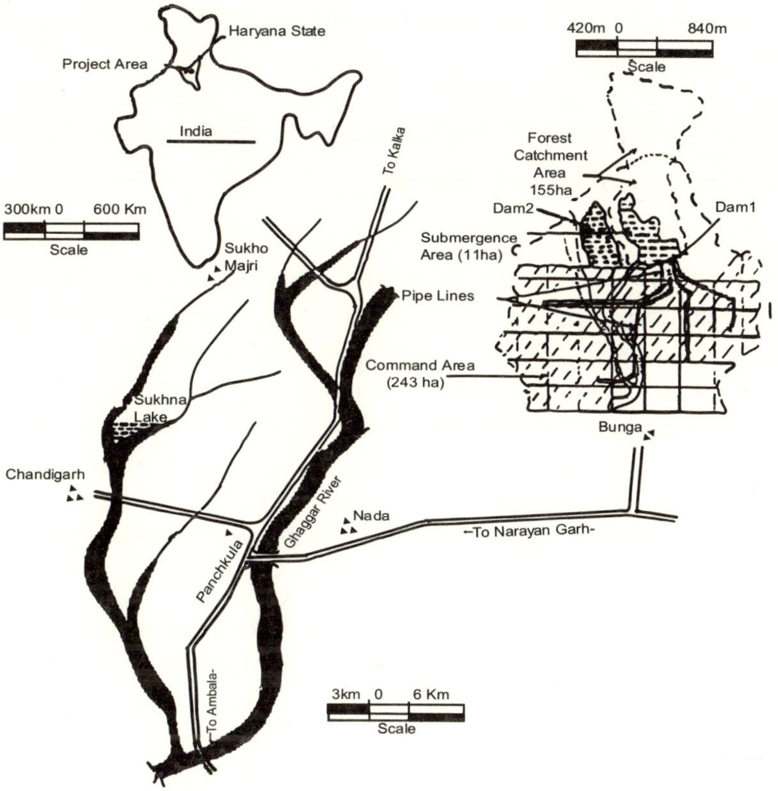

In addition to making water available for irrigation, the other important outputs of these dams have been—the almost complete elimination of floods that used to inundate this village due to heavy run-off of rainwater during the rainy season; reduction in soil erosion from the barren hill slopes, and improved availability of drinking water in the village for human beings as also for the livestock.

While the process of development and restoration of environmental degradation in the village began with the construction of water harvesting structures and setting up of water rights for village households, however subsequent institutional and other changes that occurred in the process combined together have been responsible for the transformation witnessed in this village. Therefore, while it is difficult to attribute the resultant observed changes in this village to the availability of irrigation water *per se*, nevertheless the availability of irrigation water has been the major driving force for the observed changes in Bunga.

Economic Impacts of Water Storage Dams in Bunga

Direct Impacts of Dams in Bunga

Changes in Agricultural Output and Milk Production

The availability of irrigation water has brought about significant shifts in the economy of Bunga village. The data presented in Table 7.2 and depicted in Figure 7.2 shows that the availability of water from irrigation structures has led to significant shifts in cropping patterns, a very substantial increase in crop yields, and as a result, a significant shift in crop output, specially of foodgrains and fodder crops. During the period between 1983-84 and 2001-02, the value of wheat production increased five fold (measured at 2001-02 prices), that of maize three times and that of fodder by about 3.5 times.

The available data also shows that there have been significant shifts both in the livestock number as also in the composition of livestock between the pre- and post-project period. Thus while the number of goats in the village declined from 2174 in 1983-84 to almost nil in 2001-2002 , the number of buffaloes during the period increased from 206 to an estimated 496 (an increase of 141 per cent). As a result of the increased availability of fodder in the village, reduction in migration of

Table 7.2

Changes in Agricultural Production in Bunga

	Area under Crops (ha)		*Value of Prodn. ('000 Rs.)*	
	Before (1983)	*After (2001-02)*	*Before (1983)*	*After (2001-02)*
Wheat	88	170	496.5	2421.2
Maize	80	119	332.5	1007.2
Fodder	46	82	327.7	1102
Other Agr.	Na	85	1700	1366.4

Source: Arya and Sanra (2001) and Author's calculations.

Figure 7.2

Changes in Area under Important Crops (Ha.)

Source: Table 7.2

animals to outside village and changes in livestock composition, the productivity and production of milk increased in Bunga. The annual milk production in Bunga increased from about 231 thousand litres in 1983-84 to more than 600 thousand litres in the year 2001-02.

The Bunga society, Hill Resource Management Society (HRMS), also leases out fishing rights of the check dams to a contractor on an annual lease basis. During the period 1986-1998, the Bunga society earned on an average about Rs. 4,500 per year from the sale of fish culture. During 2001-02 it earned about Rs. 12,000 from fish culture. The income from the sale of fish is credited to the society for other developmental activities.

Household Incomes

The ultimate effect of development and management of common and private resources gets reflected in income, consumption and savings patterns at the household level. The construction of the dam and availability of irrigation water have benefited all the households in the village though to varying extents. The availability of water has not only benefited agricultural production directly but also animal husbandry indirectly through the availability of a larger quantity of green fodder, dry fodder and concentrates. The benefits of livestock occurred both via changes in the composition of livestock and through the better milk yield of milch animals. These changes have been accompanied by more intensive integration of the Bunga village economy with both the economy outside the village but within the region as also with the rest of the world. The net result of these changes has been an increase in income of all sections of society.

During the year 2001-02 estimated average income per household per year in the village was Rs. 58,488 though it varied substantially between different classes of households (Table 7.3 and Figure 7.3). The income of the non-agricultural households *viz.* workers was the lowest. The income per household increased with an increase in the size of the landholding class. The average per capita income is estimated to be Rs. 8,012 with substantial income differentials across different classes of households (Figure 7.4). The average per capita income of Bunga is however still about 38 per cent of the average per capita income of Haryana State (during the year 1999-2000 per capita income in Haryana was Rs. 21,250).

An important point to observe is that the benefits of water availability have been shared almost equally by all the land owning size groups of farmers. The realised average income from agriculture per hectare of gross cropped area is almost equal in marginal, medium and large farms though in the case of small size group of farmers the average income per hectare of gross cropped area from agriculture is somewhat lower.

Table 7.3

Income Per Household (HH) and Per Unit of Gross Cropped Area (GCA): Bunga 2001-02

Farmer Category/ Workers	*Size Group (ha)*	*No. of HH*	*GCA (ha)*	*Income/ HH—All Sources (Rs)*	*Income/ Capita—All Sources*	*Income/ GCA—All Sources (Rs)*	*Agr. Income/ GCA (Rs)*
Marginal	< 1ha	38	41.2	31726	5566	29283	10415
Small	1-2 ha	54	184.1	55824	6568	16373	6235
Medium	2-3 ha	30	128.2	87880	12554	20566	10824
Large	>3 ha	18	126.8	139800	11650	19850	10047
Workers		38	0	27316	4878		
Total		178	480.3	58488	8012		

Source: Author's Survey.

Figure 7.3

Income per Household from All Sources (Rs.)

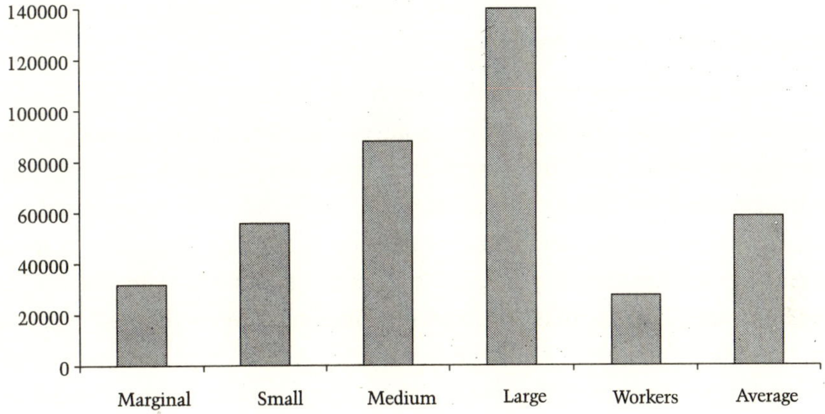

Figure 7.4

Income Per Capita from All Sources (Rs.)

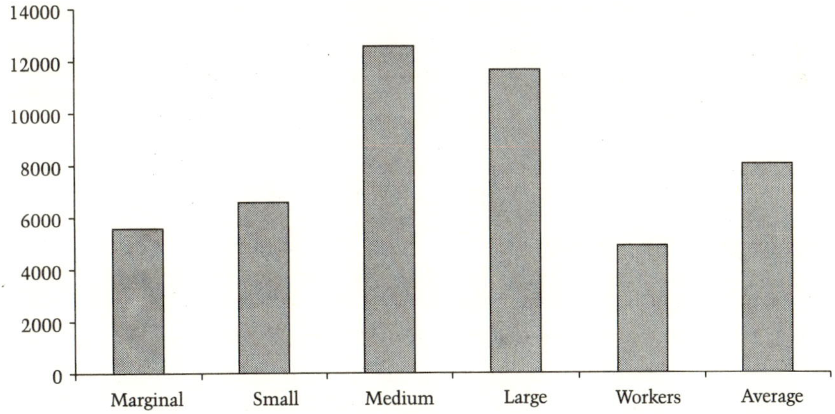

Contribution of Different Sources to Household Incomes in Bunga

As mentioned above, apart from the direct benefits of increased agricultural production, the villagers also benefited from larger milk production. While the agriculture production benefits to the land owning households from irrigation water occurred primarily due to increased crop productivity, the non-land owning households benefited from larger employment opportunities both on-farm and off-farm in such activities as construction etc. All the households however benefited from animal husbandry.

An examination of the contribution of different sources to the total household income suggests that on an average about 43 per cent of the household income in Bunga came from agriculture and 33 per cent from animal husbandry (Table 7.4 and Figure 7.5). In the case of marginal and small farmers however the larger proportion of income (between 46 to 57 per cent) was derived from animal husbandry sector while in the case of medium and large farmers, agriculture contributed more than 50 per cent to household income. The non-land owning households however got about 20 per cent of their income from employment in the agriculture sector and 17 per cent from animal husbandry. More than 47 per cent of their income came from employment in non-agriculture sectors.

Household Savings

Except for the non-land owning households, all the other households saved a part of their income. The average saving rate of the village was quite high—at about 40 per cent (Table 7.4). This is much higher than the average saving rate of Haryana state.

Indirect Impacts of Dams in Bunga

Recharge of Groundwater

In the case of Bunga village, the water table before the construction of dam was about 400 feet below the ground surface. However as a result of the seepage of water from the dam the groundwater got recharged and the water table has now risen to between 300 feet. As a result of rise in water table there has been some private investment in installation of tubewells in Bunga village. In all, about seven private tubewells have come up in the village supplying supplemental irrigation water to owners who also sell tubewell water to fellow farmers to enable them meet their additional irrigation water requirement. A tubewell exclusively for drinking water has come up in the village that has considerably improved the drinking water supply in the village.

Table 7.4

Contribution of Different Sources to Total Income and Savings of Households

Category	Size Group (ha)	No. of HH	GCA (ha)	Income from Different Sources ('000 Rs.)					Savings ('000 Rs.)	% Savings
				Agriculture	Animal Husbandry	Other Sources	Remittances	Total		
Margin	< 1	38	41.2	429 (35.6)	694 (57.4)	31 (2.6)	53 (4.4)	1206 (100.0)	193	16.0
Small	1-2	54	184.1	1148 (38.1)	1374 (45.6)	492 (16.3)	0.0	3015 (100.0)	1096	36.4
Medium	2-3	30	128.2	1388 (52.6)	641 (24.3)	608 (23.1)	0.0	2636 (100.0)	1350	51.2
Large	>3	18	126.8	1274 (50.7)	547 (21.7)	649 (25.8)	46 (1.8)	2516 (100.0)	1491	59.2
Worker		38	0.0	211 (20.4)	179 (17.2)	492 (47.4)	156 (15.)	1038 (100.0)	-15	-1.4
Total		178	480.3	4450 (42.7)	3435 (33.0)	2272 (21.8)	255 (2.5)	10411 (100.0)	4115	39.5

Source: Author's Survey.

Figure 7.5

Contribution of Different Sources to Total Income of Households, 2001-02 (Per cent)

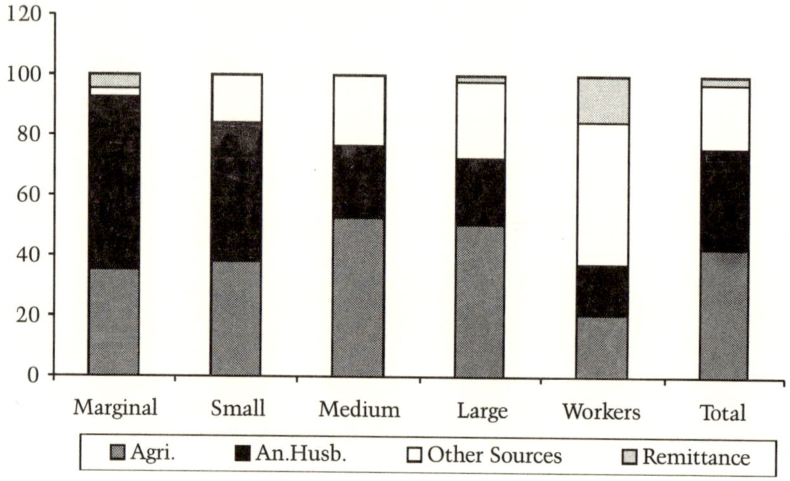

Improvement in Living Conditions

The living conditions of households in Bunga have also improved considerably. By 1992, the covered area with mud roof, wall and floor reduced by 388 per cent and modern housing area increased by 3938 per cent (Arya and Samra, 2001). Similarly area under boundary with mud wall and floor decreased by 103 per cent whereas with brick wall and mud floor increased by 817 per cent. Similarly area under semi-modern cattle shed increased by 208 per cent over the traditional design.

Integration of Village Economy with 'Outside' World

One of the important indirect impacts of the dam has been the integration of the village with the rest of India. The marketable surplus of agricultural commodities and milk production generated in the village after the construction of the dam has helped integrate the economy of the village with the rest of world. Every day almost four to five tonnes of milk are sold outside the village. A number of village youths now go out for work to the neighbouring cities of Panchkula and Chandigarh.

Multiplier Effects of the Bunga Check Dams

The Approach and Methodology

As in the case of most of the village economies of the developing world, the structure of the economic activity of the study village Bunga is also reasonably simple. Before the construction of check dams in Village Bunga, the production and consumption structure tended to be very simple. The economy of the village was more or less a subsistence economy with minimal interactions amongst production units themselves or between the village and the rest of the world. That is, however, not to imply the complete isolation of the village from the rest of the economy. The subsistence food economy co-existed with a small commercial food and non-food sector. Crop production and livestock rearing formed the core of economic activity in the village. Most of the production in this small village used to be consumed by the production units themselves and thus did not enter the exchange economy inside or outside of the village. Many of the inputs were supplied by the same production units in the form of family labour, draft power or seeds saved from previous crops etc. However all the households were neither uniform in their resource endowments nor were they self sufficient in all respects.

After the construction of the dams, the production structure became more complex and diversified, production-consumption interactions became more apparent, village income increased and interactions with the outside world increased both on the input and output sides. The list of goods and factors demanded and supplied by the village households also expanded[1]. These interactions have resulted in creating multiplier effects via linkages in production and linkages induced by higher income and consumption of households. Thus the total effects (direct and indirect) of production activities resulting from irrigation available from small check dams have been much higher than the apparent direct effects.

1. The role of outside and inside institutions in utilisation of common property resources and in shaping the development of the village increased. National policies, especially of minimum support prices of agricultural commodities, became increasing important in taking farm level decisions. However, given the small size of the village, its demand and supply has had no effect on the prices of tradables or their production.

Multipliers are summary measures that reflect the total effects of a project in relation to its direct effects. A multiplier of 1.40, for instance shows that for every one dollar of value-added generated directly by the project at maturity, another 40 cents are generated in the form of indirect or downstream effects. Thus, a multiplier is a ratio of the total effects (direct and indirect) of a dam project to its direct effects.

Estimation of multipliers requires careful analysis of the direct effects of a dam. This involves the quantification and valuation of major outputs of the dam and the assessment of the share of direct effects that are attributable to the dam project. In addition to quantification of direct impacts, it is important to capture the indirect impacts of increases in crop production and incomes resulting from water available for irrigation. These indirect impacts include purchase of fertilsers, seeds and other inputs; water sales; animal husbandry, milk production and processing; sales of consumer goods and consumer durables such as radios, televisions, bicycles, scooters, motor cycles, pumps, diesel engines etc. Due to the non-availability of these facilities within the village, most of these transactions take place in the neighbourhood area of the village where such facilities for marketing of inputs, outputs and other goods and services exist. It is therefore, important to capture these indirect impacts of crop production and higher incomes by defining an 'economic region' of the village that includes an area outside the village but within a reasonable distance from the village. Based on discussions with the village local people and through personal close observation of the prevailing situation, for the Bunga village such an 'economic region' has been defined as an area lying within a radius of about 30 kilometres of the village. Most of the strong economic ties, which the village has with the outside world, in terms of the sale of agricultural commodities, sale of milk and its processing, purchase of inputs and consumption goods, opportunities for employment etc. lie within this region.

For estimating the multplier value of the Bunga check dams, as shown in Table 7.5, for the numerator, we need to estimate the aggregate value-added (for the village and the surrounding area) under the 'With Project' situation as well as the aggregate regional value added under the 'Without Project' situation. For the denominator, we need to estimate the value added from the sectors that are directly affected by the major outputs of the dam (namely agricultural output, water supply etc.). As discussed later in this section, the estimates of the aggregate value-added under the 'With Project' situation and the 'Without Project' situation have been obtained from the use of a Social Accounting Matrix (SAM) for the 'region' defined to include the village and its neighbouring areas in Haryana state. Value-added of sectors directly affected by the project (agricultural and water sales) have also been estimated from the SAM model. The values of agricultural output and water sectors under

Table 7.5

Values of Variables Required for the Estimation of a Project Multiplier of the Bunga Check Dam Project, India

Definition of multiplier	Aggregate value-added With Project	–	Aggregate value-added Without Project
	Value-added from Crops and water Sales With Project	–	Value-added of crops and water sales Without Project

the 'without project' situation have been estimated from the data on area and yield of major crops in the pre-dam situation.

To capture the complex interlinkages among village production activities, village institutions and the outside world, village-level SAM provides a useful starting point for village wide economic analysis (Taylor and Adelman, 1996). A village SAM summarises and neatly illustrates the flow of inputs, outputs and incomes between different production activities and village households and institutions, the channeling of these incomes into consumption and investments and the exchange of goods and factors between the village and the rest of the world.

The SAM framework is a form of double entry accounting. The SAM presents income and product accounts and input-output production accounts as debit and credit entries in income balance sheets of institutions and activities. The activities in the case of Bunga village SAM include agriculture—appropriately disaggregated, animal husbandry, forestry and fisheries, and water, institutions, capital account and the rest of the world.

Entries in SAM include intermediate input demands between production sectors, income (value-added) paid by production sectors to factors of production (land, labour and capital), and the distribution of household groups' expenditure across savings, consumption of goods and services and imports from the rest of the world. The institution, HRMS, account collects water tariffs from households, income from leasing out of fisheries rights and income from forestry and redirects this income within the system on various developmental activities or saves it.

The analysis has been done using data for 2001-02. For estimating the multiplier values, the 2001-02 'Without Project' values of major production activities have been estimated on the basis of values of underlying variables as prevailing in 1983-84. It has been assumed that in the absence of the project while the cropping pattern would not have undergone significant changes, the crop yields would have nevertheless improved somewhat.

A SAM-based, fixed price, multiplier model has been used to estimate the project multiplier using a Social Accounting Matrix for Bunga for the year 2001-02. The aspects of the dam that have been analysed include: changes in the area irrigated; changes in yields and production technology that would have been unlikely in its absence. The analysis is envisaged to capture the main effects of the dam during a typical year during its lifetime, that is, say during 2001-02. The effects have been divided into direct and indirect. The indicators, capturing the effects, include production, trade, as well as disaggregated household incomes and their distribution.

The institutions in SAM include households of different types in the village—marginal farmers (operating less than 1 ha of land), small farmers (operating between 1-2 hectares of land), medium farmers (operating 2-3 hectares of land) and large farmers (operating 3 hectares or more of land); workers (landless) providing agricultural labour to land owning households and also earning their livelihood through employment in other vocations; the village society—HRMS , shops and other establishments located within the village; shops and establishments located outside the village but within the economic region of the village. As already discussed, the economic region of the village has been defined as an area lying within a radius of about 30 kilometres of the village within which the village has strong economic ties in terms of the sale of commodities, purchase of inputs and consumption goods and also for earning wage employment. There are nine production sectors. The data source for estimating the SAM is a recent study of the Bunga economy.[2] The study collected detailed data from 47 sampled households (there are 178 households in the village) selected according to an appropriate stratified sampling scheme. The study also reports data collected from shops and other institutions.

The aggregated version of the Bunga SAM is presented in Table 7.6. The key features of the economy of Bunga are analysed with the help of the aggregated SAM. In 2001-02, the gross value of output in Bunga was Indian Rupees (Rs.) 20,065 thousand and the value added was Rs. 9,373 thousand (about 47 per cent of the total production). Gross value of agricultural output (including water) was Rs. 6099.7 thousand (30.4 per

Table 7.6

An Aggregated Social Accounting Matrix for Bunga 2001-02 ('000 Rs.)

	HHs	*Agriculture*	*Animal Husband.*	*Forest & Fish*	*Shop-in*	*Shop-out*	*HRMS*	*Capital*	*ROW*	*Total*
HHs	0	4415.4	3256.4	0	26.7	1575.5	0	0	98.9	9372.9
Agriculture	1148.7	497.6	2681.5	0	629.4	883.5	50	209	0	6099.7
Animal Husbandry	1696.8	412.4	0	0	249.6	2468.8	0	1079.4	0	6117
Forest & Fish	0	0	0	0	0	0	0	59	0	59
Shop-in	744.5	537.6	0	0	0	226.6	0	0	0	1678
Shop-out	1653.4	0	0	0	234.7	0	0	0	3549.8	6111.3
HRMS	0	25.4	0	59	0	0	0	0	0	84.4
Capital	4129.5	0	0	0	447.1	555.6	34.4	0	-3323.7	1828.2
ROW	0	0	0	0	0	0	0	480.8	0	480.8
Total	9372.9	6099.7	6117	59	1678	6111.3	84.4	1828.2	480.8	

2. *Collective Action and Income Distribution Aspects of Water Management; A Case Study of the Bunga Check Dam, Haryana, India.* Draft Report prepared by Meera Bhatia. January 2003.

cent of the total value of production) while output of animal husbandry was Rs. 6117 thousand (about 30.5 per cent of total value of production).

A SAM-based Multiplier Model of Bunga Village and Neighbouring Economic Region

This study uses a SAM-based, fixed price, multiplier model which has been calibrated for the year 2001-2002.

Since the SAM is a double-entry accounting system, we can use either the row or column accounts for its presentation. The rows provide the statement of receipts for each account, whereas the columns provide the statement of expenditure. For example, the row for the *k* th type of household represents the receipts of household *k*, i.e.

Total income of household k = Value-Added received by household k from the production sectors + Income transfers from outside the region to the household.

The corresponding column for the household provides the details of the expenditures of the household, i.e.

Total expenditure = Expenditure on domestic goods + Savings + Taxes + Expenditure on imported goods.

Table 7.7

A SAM-based Multiplier Model of the Bunga Regional Economy

	Expenditures				
Receipts	*Households (HHs) k=1,..,6*	*Production Sectors i=1,2,...9*	*Capital Account*	*Rest of the World*	*TOTAL*
Households k=1,..,6	0	$\sum_i W_{ki}X_i$	0	R_k	Y_k
Production Sectors i=1,2,...9	$\sum_k c_{ik}Y_k$	$\sum_j a_{ij}X_j$	I_i	E_i	X_i
Capital Account	S_k	S_i	0	S_R	F
Rest of the World	M_k	M_i	0	0	M
TOTAL	Y_k	X_i	F	M	

Table 7.7 shows the interactions among different accounts in the SAM. These interrelationships are explained in more detail in the following equations.

In the SAM based model there are 9 production sectors including wheat, maize, fodder, other agriculture, animal husbandry, water, fishing & forestry, shops within the village and shops outside the village. As mentioned before, there are 5 types of households namely marginal farmers, small farmers, medium farmers, large farmers and workers. There is also an institution namely the HRMS. The notation used in the following is as follows:

The subscript $k = 1,2,\ldots 6$ denotes the 5 types of households and the HRMS, while i, j denote the production sectors.

Y_k = total income of household/ institution of type k

X_i = value of gross output of sector i

w_{ki} = coefficient (ratio) of value-added received by household type k to output of sector i

R_k = income transfers from abroad to households of type k

C_{ik} = proportion of income of household k that is spent on the purchases of sector i's output

a_{ij} = input-output coefficient for production sectors

J_i = investment demand of sector i 's output

E_i = exports of sector i 's output

F = total savings in the village economy

S_k = total household savings

S_i = savings/investment by production sectors

S_R = exogenous inflow of capital

M = total value of imports

M_k = expenditure on imported goods by household of type k

M_i = value of imported intermediates purchased by sector i

As in a Leontief system, we assume that all the structural relations (both behavioural and technological) are linear or at least they can be approximated to linear functions. A description of the various rows of the SAM-based multiplier model are presented in the following equations.

Households:

$$Y_k = \sum_i W_{ki} X_i + R_k \text{ for } k = 1,2,\ldots 6.\ i = 1,2,\ldots 9.$$

[Total income of households/institution of type k] = [Total factor payments received from all production sectors] + [income transfers from abroad to households/institution of type k]

Production Sectors:

$$X_i = \sum_k C_{ik} Y_k + \sum_j a_{ij} X_j + J_i + E_i$$

[Gross Output of production sector i] = [Sum of all households'/ institution's demands for output of sector i] + [Intermediate demand by sector j for output of sector i] + [investment demand] + [Exports of sector i]

Capital Account Receipts:

$$F = \sum_k S_k + \sum_i S_i + S_R$$

[Total Savings] = [Total household/institution savings] + [Savings/ Investment by production sectors]+ [exogenous inflow of capital]

Rest of the World Account:

$$M = \sum_k M_k + \sum_i M_i$$

[Total Imports] = [expenditure on imported goods by all households/institutions] + [value of imported intermediates purchased by production sectors]

Simulations under the 'Without Project' Situation

Bunga check dams have contributed significantly to the increase in the output of agricultural commodities and milk production and these in turn have been an important source of growth in the village economy. These increases have inevitably generated downstream growth in many other sectors of the village economy. In addition, there have been some other minor sources of growth that are not related primarily to the production of agricultural crops and milk. These include small government expenditures in infrastructure, education and health. In order to unravel the impact of increases in agricultural commodities and milk production from all the other autonomous sources of growth in the economy, it is necessary to construct a picture of the 'regional economy' as it would have been in 2001-02 had the Bunga check dam not been constructed. A key element in the analysis is to construct an estimate of this hypothetical situation for the economy and then to make comparison against the situation with the construction of the dam.

We have had access to some of the actual data on some of the relevant variables under the 'With Project' situation with their counterparts in the 'Without Project' case. This set of variables comprises all the elements of a SAM for the 'region' in each situation, assuming fixed prices.

Bunga Village Economic Region—'Without Project' Scenario

Cropping Pattern

Before the construction of the project, in the year 1983-84 the cropping pattern of the village was as follows (Table 7.8)

The entire area under all the crops was unirrigated. The most important contribution of the project has been that the availability of irrigation has enabled cultivation of wheat on a much larger scale and

Table 7.8

Cropping Pattern and Crop Yields in Bunga: 1983-84

Crop	*Area in Ha*	*Per cent Share*	*Crop Yield (kg/ha)*
Rabi Season Crops			
Wheat	88	16.0	910
Gram	65	11.8	390
Arhar	72	13.1	702
Other *Rabi* crops	18	3.3	702
Kharif Season Crops			
Maize	80	14.6	875
Groundnut	72	13.1	665
Bajra	93	16.9	1050
Jowar	46	8.4	950
Other *Kharif* crops	15	2.8	1050

Source: Arya and Samra (2001).

under irrigated conditions. In addition, the availability of irrigation has enabled cultivation of Berseem fodder that was not being cultivated in unirrigated conditions and a reduction in fallow area.

It is safe to envisage that in the likely scenario of the project not having been built, the area under cultivation in the year 2001-02 would have remained at the same level as prevailing in 1983-84.

The yields of some of the crops prevailing during 1983-84 in the Village Bunga were as given in Table 7.8. During the intervening period 1983-84 to 2001-02 it is very likely that the crop yields would have risen somewhat even if the project had not been undertaken and the crops continued to be cultivated under rainfed conditions. We assume that the crop yields would have risen by about 10 per cent during the intervening period. Using the crop prices prevailing in 2001-02, the likely gross value of agricultural production in the 'without project' scenario would have been as follows (Table 7.9)

Table 7.9

Value of Agricultural Output in Bunga under Without Project Conditions: 2001-02

Crop	*Value ('000 Rs.)*
Wheat	546
Maize	366
Fodder	361
Other Agriculture	1648
Total	2921

Source: Author's Calculations.

A comparison of the value of crop output under the 'with' and 'without project' scenarios in the year 2001-02 in Bunga suggest that the value of all the three important crops—wheat, maize and fodder increased significantly as a result of the project (Table 7.10, Figure 7.6). The gross value of output of the four crops increased by almost 102 per cent.

Results of Multiplier Analysis

Table 7.11 presents the results of the estimate of value-added multipliers based on the methodology and assumptions described above.

Table 7.10

Value of Crop Output in Bunga: With and Without Project—2001-02

Crop	*Value ('000 Rs) 2001-02 With Project*	*Value ('000 Rs) 2001-02 Without Project*
Wheat	2421	546
Maize	1007	366
Fodder	1102	361
Other Agriculture	1366	1648
Total	5896	2921

Source: Author's Calculations.

Figure 7.6

Value of Crop Output With and Without Project (Rs.'000)

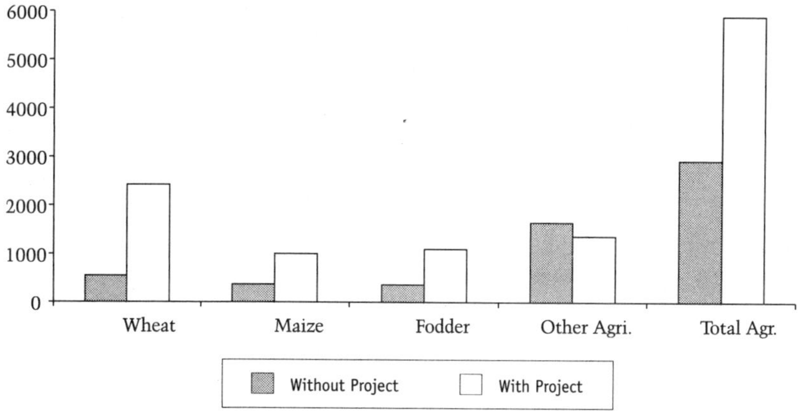

Table 7.11

Estimated Value for Multiplier Effects of the Bunga Check Dam, India

Definition of Multiplier	$\dfrac{\text{Direct Effects} + \text{Indirect Effects}}{\text{(Direct Effects)}}$
Value (Thousands of Indian Rupees (Rs.)	$\dfrac{\text{Aggregate Value Added (With Project)} - \text{Aggregate Value Added ('Without Project')}}{\text{Value Added of Sectors Affected Directly ('With Project')} - \text{Value Added of Sectors Affected Directly ('Without Project)}}$ $\dfrac{10241 - 6759}{4711 - 2235} = 1.41$

Notes: Sectors affected directly by the Dam are: Foodgrains, fodder, water, fishing and forestry.

Source: See text and tables.

The results show that the regional value-added under the 'With Project' scenario at Rs. 10,241 thousand is larger than the regional value added under the 'Without Project' scenario by Rs. 3,482 thousand. Compared with this, the value-added from sectors affected directly by the project at Rs. 4,711 thousand was larger than the corresponding value-added under the 'Without Project' scenario by Rs. 2,476 thousand. Thus, in the aggregate, the project induced an increase of Rs. 3482 thousand in regional value added. Of this, Rs. 2,476 is due to an increase in the output of sectors directly affected by the Bunga dam.

This gives a multiplier value of 1.41 i.e. Rs. 3,482/2,476. Thus, for every rupee (100 paise) of additional value-added directly by the project in the agricultural sector, another Re 0.41 (41 paise) were generated in the form of downstream or indirect effects.

Income Distribution Impacts of the Bunga Check Dam

The income distribution impacts of dams indicate that while all the sections of the households have benefited from the construction of the dam, the benefits have accrued relatively more to the households having land as compared to landless households. However, even within the land holding households the magnitude of income impact has differed somewhat—the relatively smaller farmers have benefited more as compared to the large farmers.

As discussed earlier, there are various categories of households in the village mainly distinguished by their land ownership. These are:

(i) Marginal farmers (operating on less than 1 ha of land) account for 21.3 per cent of total households.

(ii) Small farmers (operating between 1-2 hectares of land) account for 30.4 per cent of total households.

(iii) Medium farmers (operating between 2-3 hectares of land) account for 17 per cent of total households.

(iv) Large farmers (operating between 3 hectares or more of land) account for 10 per cent of total households.

(v) Workers (landless) providing agricultural labour to land owning households and also earning their livelihood through employment in other vocations inside or outside the village (accounting for 21.3 per cent of total households).

The income distribution impacts of the Bunga dam have been analysed in three ways:

(i) By comparing the shares of marginal and small farmers, workers and other households in aggregate incomes under the 'With' and 'Without Project'.

(ii) By comparing the differences in aggregate income levels of various household categories under the two scenarios of 'Without Project' and 'With Project'.

(iii) By assessing direct and indirect components of income differences under the 'With' and 'Without Project' scenario.

Shares of Small and Marginal Farmers and Workers in Aggregate Incomes under the 'With' and 'Without Project' Scenarios

The SAM-based, fixed price, multiplier models have been used to assess the income effects of additional irrigation provided by the Bunga dam. This has been done for each of the household categories described above. Table 7.12 and Figures 7.7 and 7.8 present results of the shares of different categories of households in the aggregate rural income under the 'With' and 'Without Project' situations. For example, for the marginal farmer households (which account for 21.3 per cent of households), the share of their income in the aggregate rural income levels is 11.5 per cent under 'Without Project' situation compared with 11.6 per cent under the 'With Project' situation. For small farmers, accounting for 30.4 per cent of total households, the share of income at 29 per cent of the total under 'With Project' situation is higher than 27 per cent under 'Without Project' situation. However, when a comparison is made between shares of large farmers under the 'With' and 'Without' situations, the share of large farmers (10 per cent of total households) is marginally lower under the 'With Project' situation. Similarly, although worker households (21.4 per cent of total households) with an income level of Rs. 842 thousand under the 'Without Project' situation increased their income levels to Rs. 1038 thousand, their share in the total is lower (10 per cent) under the 'With Project' than under 'Without Project' situation (12 per cent of the total).

Figure 7.7

Share of Income of Different Classes of Households—Without Project: Percentages

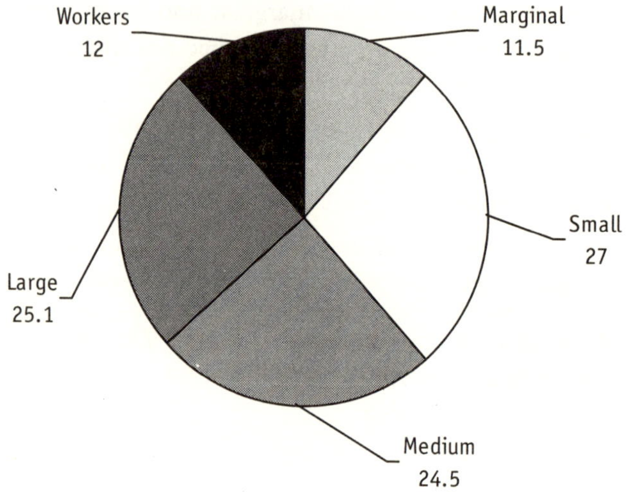

Figure 7.8

Share of Income of Different Classes of Households—With Project: Percentages

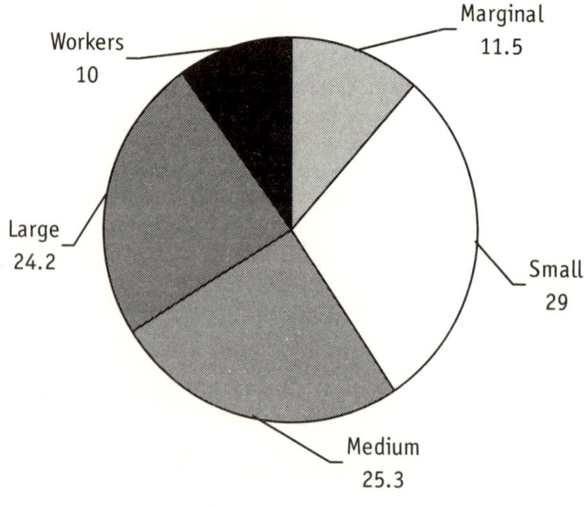

Table 7.12

Income Shares With and Without Project (In Indian Rs.)

Category of Household	*Without Project*	*Per cent of Total*	*With Project*	*Per cent of Total*
Marginal Farmers	803	11.5	1206	11.6
Small Farmers	1891	27.0	3015	29.0
Medium Farmers	1715	24.5	2636	25.3
Large Farmers	1762	25.1	2516	24.2
Workers	842	12.0	1038	10.0
Total	7013		10411	

Source: Simulations using SAM for Bunga. See Text

Differences in Aggregate Income Levels of Various Household Categories under the 'Without Project' and 'With Project'

Table 7.13 and Figure 7.9 show that all households have higher income levels under the 'With Project' situation than under the 'Without Project' situation. For example, for the marginal farmers, the difference in income levels (under the 'With' and 'Without project' situations) are estimated to be Rs. 403 thousand. Their income under 'With Project' situation is 50 per cent higher over the income level under the 'Without Project' situation (Rs. 803 thousand). In the case of small farmers, the

Table 7.13

Differences in Income Levels of Various Household Categories 'With Project' and 'Without Project' Situations (Income in '000 Rs. from All Sources)

Category of Households	*With Project 2001-02*	*Without Project 2001-02*	*Percentage Difference*
Marginal Farmers (less than 1 Ha)	1206	803	50.1
Small Farmers (1-2 Ha)	3015	1891	59.4
Medium Farmers (2-3 Ha)	2636	1715	53.7
Large Farmers (> 3 Ha)	2516	1762	42.8
Workers	1038	842	23.3
Total	10411	7013	48.5

Source: Simulations using SAM for Bunga. See text.

difference in income is Rs. 1124 thousand, which shows that their income under the 'With Project' situation is 59 per cent higher than the

Figure 7.9

Income of Households With and Without Project (Rs. '000)

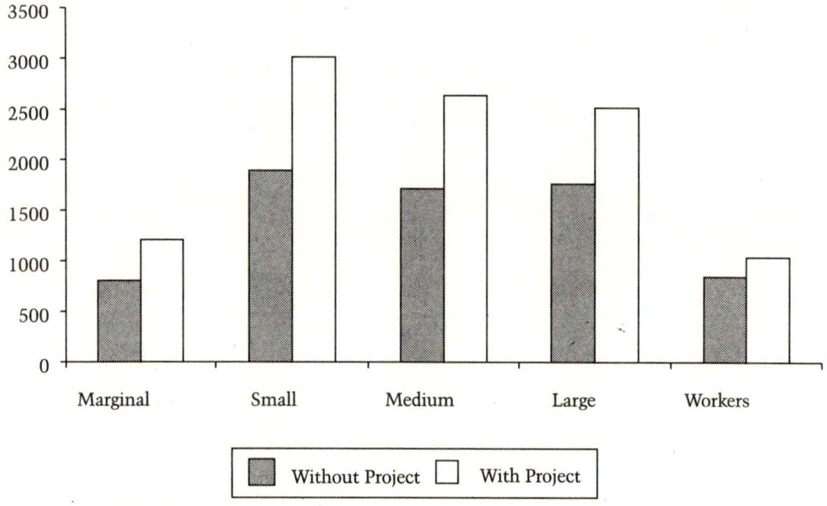

income level of Rs. 1891 thousand under the 'Without Project' situation.

However, in the case of workers, the difference in the income levels under the two situations is relatively low, only 23.3 per cent. The level of income of workers households under the 'Without Project' situation is Rs. 842 thousand which increases to Rs. 1038 thousand under the 'With Project' situation. The results show that the gain for land holding households (except for large farmers) from the dam is relatively higher than the average difference of 48 per cent between village-level incomes under the 'With' and Without Project' situations. Further, the worker households do not benefit much from the dam because the demand for increased labour from irrigated crops has been met from family labour by most of the farm families. Due to relatively small size of farm holdings, the farm households do not increase their demand for hired labour that would have benefited landless households. However, the worker households benefit indirectly from the dam in terms of higher incomes from milk production and shops (as discussed later in this section).

Direct and Indirect Components of Income Differences under 'With' and 'Without Project'

The SAM-based multiplier models have also been used to estimate the direct and indirect income effects of additional irrigation provided by the Bunga dam. This has been done for each of the household categories, namely, marginal, small, medium and large farmers and workers. The SAM coefficients for each household and each production sector have been used to estimate the value-added received by each household category from each sector. These SAM coefficients have been used to estimate the component of a household's income that is due to value-added from sectors affected directly by the output of the dam, namely irrigation and water supply. Similarly, SAM coefficients for the remaining sectors have been used to estimate the component of income that is due to value added from sectors affected indirectly by the output of the dam. This type of analysis is based on the usual assumptions for this type of fixed-price SAM-based multiplier model namely that the value-added per unit of output remains the same for each sector and the adjustments take place through changes in quantity and so on.

The procedure used to estimate the share of direct and indirect components in the household incomes is as follows. First, the SAM coefficients have been used to estimate values of a new SAM for the 'Without Project' situation by fixing the outputs of sectors directly affected by the dam (agricultural sectors). Then the difference between the values of outputs and household incomes in the two SAMs are used to estimate differences in incomes under the 'With Project' and 'Without Project' situations. The difference in the household incomes is further divided into two components—one from the direct sectors and the other from the indirect sectors. To calculate the component from the direct sectors we look at the row of each household in the matrix of income differences and sum up the incomes the household received from the sectors directly affected by the project namely—production of wheat, maize, fodder and sales of water. To calculate the component from the indirect sectors we added the incomes received from the remaining sectors (e.g. milk production, shops).

The resulting estimates show, as predicted, that a major part of the income of rural households comes from the output of sectors directly affected by the dam. Table 7.14 and Figure 7.10 present the results of the disaggregation of the differences in household incomes under 'With' and 'Without Project' situations. For example, for the marginal and small farmers, about two-thirds of the difference between incomes under the 'With' and 'Without' situations is due to higher output and value-added in sectors that are directly affected by the dam, namely agricultural crops. The rest, about one-third, of the difference in income is attributable to changes in value of output and value-added of sectors that are affected indirectly (milk production, shops etc). In the case of medium and large farmers, as much as 77 per cent of the difference in income is due to changes in output/value-added of sectors that are directly affected by the dam. However, in the case of worker households,

Figure 7.10

Income Gains from Bunga Dam—From Direct and Indirect Sectors (Percentages)

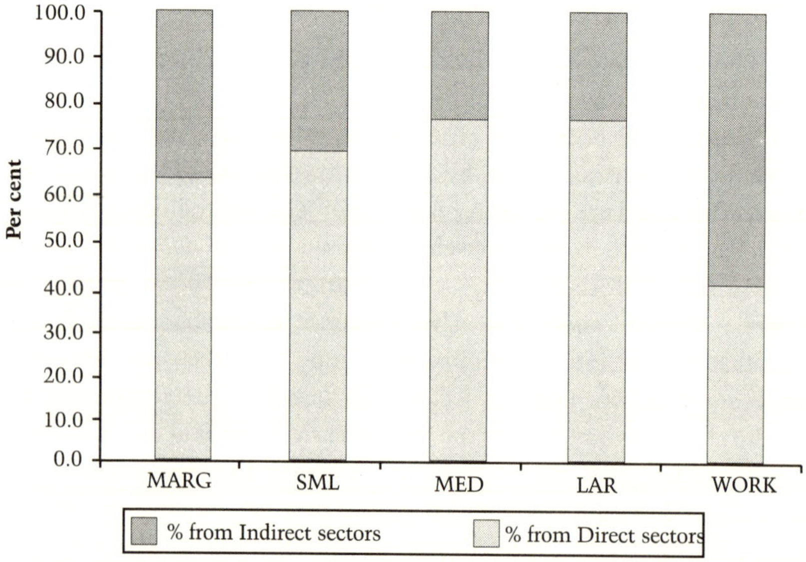

Table 7.14

Income Gains from the Bunga Dam—From Direct and Indirect Sectors

Category of HH	Income from Direct Sectors	Income from Indirect Sectors	Total Income from Direct and Indirect Sectors	Per cent from Direct Sectors	Per cent from Indirect Sectors
Marginal Farmers	255	147	402	63.4	36.6
Small Farmers	776	347	1123	69.1	30.9
Medium Farmers	704	217	921	76.5	23.5
Large Farmers	581	173	754	77.0	23.0
Workers	75	122	197	38.1	61.9

Source: Simulations Using SAM for Bunga. See text

the situation is completely reversed. As much as 62 per cent of the difference (under the 'With' and 'Without Project' situations) is accounted for by changes in outputs of sectors that are indirectly affected by the dam, namely animal husbandry, shops etc.

Economic Evaluation of Additional Investments in Check Dams in Bunga

The results obtained have clearly demonstrated the benefits that the village economy has reaped from the water made available from these dams. The benefits of these check dams that have been derived so far are, however, much less than the benefits that can potentially be derived. The Bunga village currently has two check dams but due to heavy siltation in the dams over the years their command area has declined. As a result, the dams are currently able to supply water to only wheat during the *Rabi* season. In addition, limited amount of water is also made available for meeting the water requirements for cultivating fodder. If additional water can be stored in the present dams by desilting them and/or an additional dam can be built to supplement the available water, the villagers can be assured of a second irrigated crop in the *kharif* season.

In the neighbouring areas of Haryana and Punjab the availability of irrigation during the *kharif* season has resulted in shifts in area in favour of Paddy from the other crops. In most cases this shift has been achieved at the cost of the area under maize. We do expect that if additional water were made available in Bunga during the *kharif* season a similar change in crop pattern would emerge. The water in the Bunga dam during the year 2001-02 was able to supply water for irrigating 170 hectares of wheat land. The water requirements of Paddy are however much more than that of Wheat. We assume that the water requirement of 1 hectare of paddy is roughly equivalent to 4 hectares of wheat crop. If through construction of a new dam or desilting of the existing dam the currently available water supply can be doubled and the additional water so generated is supplied in the *kharif* season, it could irrigate about 42 hectares of paddy land. We assume that this paddy land comes from the land that is currently under maize. As a result the area under maize declines from the current level of 119 hectares to 77 hectares. Taking the average yield of Paddy for Haryana State at 3.6 tonnes per hectare and the prevailing procurement price of paddy, the gross value of Paddy output from this land would be of the order of Rs. 800 thousand . The loss in gross value of output of maize on account of conversion of this land would be around Rs. 357 thousand. The net increase in gross value of agricultural output is likely to be of the order of 443 thousand. Taking the ratio of value-added to gross value of output at approximately 0.67 the net value addition is likely to be of the order of Rs. 300 thousand or Rs. 0.3 million per year.

While any official estimate of the cost of desilting and other costs are not available for the purpose of the present exercise we estimate that the cost of construction of a small dam upstream (to slow down the rate of desilting in the future years) and the cost of desiltation of the two existing dams would be around Rs. 1.5 million.[3] Assuming that O&M costs will be 5 per cent of capital cost, this gives a sum of the present values of O&M costs at Rs. 0.46 million assuming an economic life of 10 years at 10 per cent discount rate. Thus, the present value of total capital and O&M costs would be of the order of Rs. 2 million.

Taking the net additional annual value of output at Rs. 0.3 million per year, the present value of net benefits will be Rs. 1.85 million (0.3 million x 6.16) taking an economic life of 10 years at 10 per cent discount rate. This is marginally less than the present value of total costs and hence will give a benefit-cost (B/C) ratio of 0.94 or less than 1.

3. Based on discussions with knowledgeable persons in the village. The villagers had got these estimates from some contractors who do the desiltation jobs.

However, if we use the value of the multiplier to incorporate the indirect effects of the investments in the check dams, the benefit cost ratio may be different. Taking a multiplier value of 1.4, as obtained for the investments in check dams in the past and multiplying net benefits by 1.4, the net value added per year is estimated to be of the order of Rs. 420 thousand or Rs. 0.42 million. This gives a net present value of additional benefits of 2.6 million over a ten-year life at 10 per cent discount rate. If we take the multiplier value to reflect additional net benefits to the society, then the benefit-cost (B/C) ratio is equal to 1.33 or marginally higher than 1.0.

Summary and Conclusions

Bunga provides an excellent example of how creation of even small check dams, coupled with complementary resource management policies and development of appropriate institutions for their management, can transform the economy of the village and neighbouring areas. The results obtained using a Social Accounting Framework for the Bunga village suggest that the magnitude of the multiplier effect, though not very high is, still substantial.

The results of the multiplier analysis show that as a result of the check dam at Bunga the value added in the 'economic region' of the check dam under the 'With Project' scenario has been higher than the value added under the 'Without Project' scenario. In the aggregate, the project induced an increase of Rs. 3,482 thousand in 'regional' value added. Of this, Rs. 2,476 was due to increase in the outputs of sectors directly affected by the Bunga dam. This gives a multiplier value of 1.41 i.e. Rs. 3,482/2,476. Thus, for every rupee (100 paise) of additional value added directly by the project in the agricultural sector, another Re 0.41 (41 paise) were generated in the form of downstream or indirect effects.

The results on income distribution effects of investment in check dams show that small and marginal farmers (operating between 1-3 hectares of land) benefit relatively more than the marginal or large farmers. Workers do not benefit much from the dam since the demand for hired labour does not increase since farm households have enough family labour. For example, in the case of small farmers, the aggregate income under the 'With Project' situation is 59 per cent higher than the income level under the 'Without Project' situation. However, in the case of workers, the difference in the income levels under the two situations is relatively low, only 23.3 per cent. Further, 67 to 77 per cent of income gains for farm households come from the agricultural sector and a quarter of the benefits come from the indirect effects of the dam.

The multiplier effects of check dams would be higher if the indirect effects of Bunga dams outside the village (through sale of milk and animals) were included in the analysis.

Can small check dams bring about the level of economic development witnessed in Bunga at other locations as well specially outside the Himalayan Shivalik region where these villages are located? It is difficult to hazard a guess but probably would depend on a number of conditions prevailing at the pre-project stage—socioeconomic conditions, availability of leadership, the degree to which participation can de developed between and amongst different agencies and people, institutional development, role played by catalytic agencies, positive government policies etc.

References

Agarwal Anil and Sunita Narain (1999). *Community and Household Water Management: The Key to Environmental Regeneration and Poverty Alleviation.* Paper presented to EU-UNDP. Feb. Brussels Conference.

———. (1989). *Towards Green Villages—A Strategy for Environmentally Sound and Participatory Rural Development",* New Delhi: Centre for Science and Environment.

Agnihotri, Y, S.P. Mittal, S.S. Grewal and P.R. Mishra (1986). "Economic Evaluation of Water Resource Development in Shivalik Foot Hills—A Case Study", *Indian Journal of Soil Conservation* 14(2).

Anonymous (2001). "A Tribute to P.R. Mishra", *Down to Earth,* May 31. pp.52-53.

———. (1998). "Sukhomajri at the Cross Roads", *Down to Earth,* December 15. pp.30-36

Arya, Suman Lata and J.S. Samra (2001). *Socio Economic Implications Appraisal of Integrated Watershed Management Project at Bunga.* Chandigarh: Central Soil and Water Conservation Research and Training Institute Research Centre.

Bhatia, R., M. Scatasta, and R. Cestti (2003). "Study on the Multiplier Effects of Dams: Methodology Issues and Preliminary Results", Presented at the *Third World Water Forum.* Kyoto, Japan. March 16-23.

Bhatia, Meera (2003). *Collective Action and Income Distribution Aspects of Water Management; A Case study of the Bunga Check Dam.* Haryana, India. Draft Report.

CSE (1999). *State of India's Environment: The Citizen's Fifth Report*, Vol. I. New Delhi: Centre for Science and Environment.

CSWCRTI (1982). *Sedimentation in Sukhna Lake: Status Report.* Chandigarh: Central Soil and Water Conservation Research and Training Institute.

Chopra, Kanchan, Gopal K. Kadekodi and M.N. Murthy (1990). *Participatory Development: People and Common Property Resources.* New Delhi: Sage Publications.

———. (1988). "Sukhomajri and Dhamala Watersheds in Haryana: A Participatory Management", New Delhi: Institute of Economic Growth. *mimeo*

Datta, Sumana (1999). "Differing Perspectives", *Down to Earth*, April 30. pp. 56-57

Gulati, S.C. and Suresh Sharma (1998). Population-Poverty-Environment Interface: Case Study of Sukhomajri, Jattanmajri, Ralegan Siddhi and Panoli Villages". New Delhi: Centre for Science and Environment. *mimeo.*

Koppen, Barbara van (1999). *Sharing the Last Drop: Water Scarcity, Irrigation and Gendered Poverty Eradication.*

Mishra, P.R. and Madhu Sarin (1987). "Sukhomajri: A New Model of Eco-Development", *Business India.* November 16-29. pp. 78-90.

Mittal, S.P, S.S. Grewal and P.R. Mishra (1980). *Operational Research Projection Watershed Development for Sediment Drought and Flood Control—Sukhomajri.* Chandigarh: Central Soil and Water Conservation Research and Training Institute.

Rao, S.L. and I. Natarajan (1996). *Indian Market Demographics: The Consumer Classes.* Global Business Press/National Council of Applied Economic Research. New Delhi. p.49.

SPWD (1986). "Case Studies of Sukhomajri and Jawaja", Paper presented at the *International Conference on the Economics of Dryland Degradation and Rehabilitation.* Canberra. Society for the Promotion of Wasteland Development

Sarin, Madhu (1996). *Joint Forest Management: The Haryana Experience*. Ahemdabad: Centre for Environmental Education.

Seckler, D. and Deep Joshi (1980). *Sukhomajri: A Rural Development Programme in India.* New Delhi: The Ford Foundation

Sharma, Sudhirendar (2002). "Reinventing Watersheds Proposing a New Approach to India's Rural Drought Problems". *D+C Development and Cooperation* 1: 24-26, January/ February 2002.

Stewart, Willam (n.d.). "Evolving Watershed Management System in the Shivalik Hills of North West India". (mimeo).

Taylor, J. Edward and Irma Adelman (1996). *Village Economies: The Design, Estimation and Use of Village-wide Economic Models.* Cambridge University Press.

8 The High Dam at Aswan

Sherman Robinson, Ken Strzepek,
Moataz El-Said and Hans Lofgren

Introduction and Overview

The purpose of this study is to generate an improved understanding of the impact of the High Dam at Aswan on Egypt from an economy wide perspective. Our primary analytical tool is a CGE model of Egypt. The model is used to carry out comparative-static simulations of Egypt's economy with and without the Dam. The analysis covers the impact of the dam through the following channels: changes in the supplies of irrigated land and water; changes in the supplies of electric power; changes in yields and production technology (primarily changes in fertiliser use); and real costs associated with the investment relative to other investments (in flood control and hydropower) that would have been likely in its absence. In addition, the analysis considers the implications of the fact that, without the dam, the performance of Egypt's economy in each year would have depended on Nile flow levels. The model is built around a 1996/97 SAM for Egypt. For our simulations, we draw on a wide range of additional information, most importantly historical data on Nile flows and assessments of various aspects of the costs and benefits analyses of the Dam in the literature.

This analysis tries to explore the main impacts, both costs and benefits, of the dam in a typical year during its lifetime. We divide the impacts into direct and indirect. The impact indicators include production, trade, as well as disaggregated household incomes and their distribution. A virtue of the relatively simple framework (compared to a full dynamic model) is that it is relatively straightforward to implement. Our analysis provides a prototype that can be applied to other cases. The only requirement is to put together a database, the components of which we have clearly identified.

We proceed as follows. The following section provides a general background on the Dam and a review of cost-benefit analyses of the Dam, both in terms of factors considered and overall findings. Our review points to the importance of an economy wide approach in the analysis of the impact of large dams, including the High Dam in Egypt. We next present the CGE model that is used in this study and discuss the application of such models to the cost-benefit analysis of major investment projects, like dams. We also note that SAM multiplier analysis—characterised by the assumption that the prices of commodities, factors and foreign exchange all are fixed—is a special case of CGE analysis. Subsequent to this, we carry out a set of simulations under alternative assumptions about the working of the economy (with "SAM multiplier" assumptions as one extreme case). The simulations are used to assess how Egypt's economy would have performed in 1996/97 without the dam and how the economy, with and without the Dam, would have been affected by year-to-year variations in Nile flows.

The Nile, Egypt and the High Aswan Dam

The Hydrology of the Nile

The Nile is the birthplace of hydrology. Since the Nile has been of critical economic importance to both ancient and modern civilisations alike, its flow has been studied for more than 5,000 years, and recorded for more than 13 centuries. Its source has been a mystery as well as the predicting of its annual rise and fall. The Nile also exhibits cycles of years below the mean flow, drought years, followed by years above the mean, flood years. The impact of this tendency is well documented in the Book of Genesis 18 which describes the Nile's seven years of flood and seven years of drought and Joseph's management of the resulting agricultural production.

The Nile River flow entering Egypt is made up of runoff from two very different hydrological systems. One, the White Nile, and two, is the Blue Nile and Atbara. The two systems join in northern Sudan (At Khartoum). The White and the Blue Niles form the main Nile with the Atbara joining the Nile shortly downstream well before entering Egypt. Figure 8.1 shows the location of these two systems.

Figure 8.1

Nile River Basin

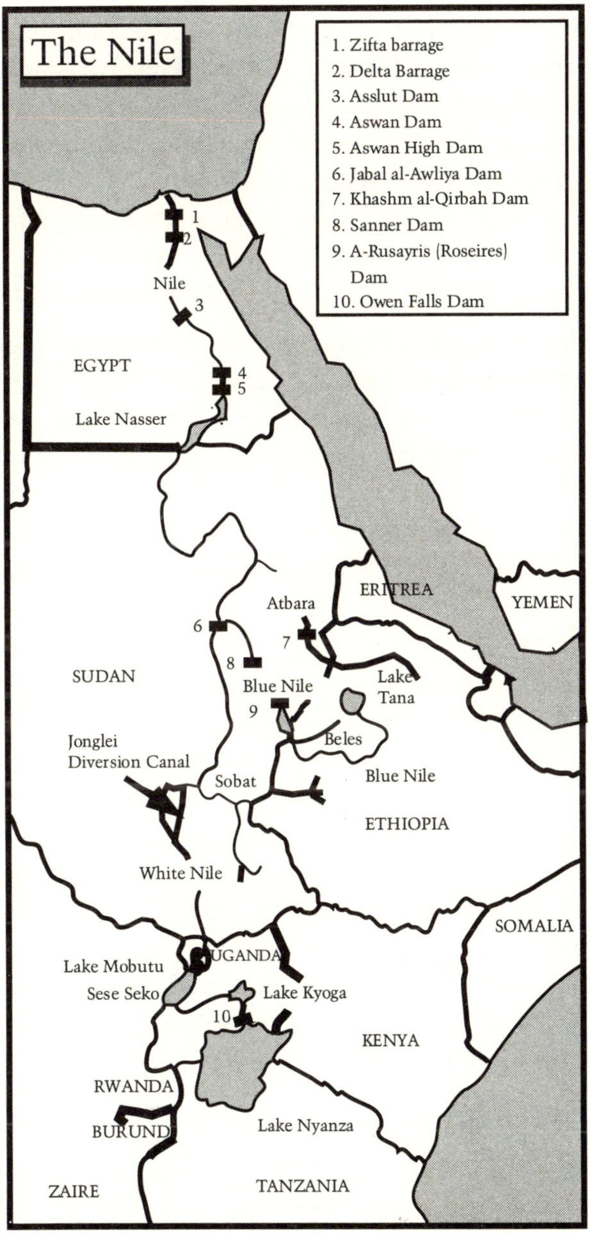

The White Nile rises in the Equatorial Lakes of DR of Congo, Tanzania, Kenya, Rwanda, Burundi, and Uganda. It is dominated by the outflow of Lake Victoria. The White Nile equatorial stream flow plummets off the Equatorial escarpment onto the plains of Sudan where together with the runoff from the Gebel el Jebel it spreads out over massive wetlands of the Southern Sudan called the Sudd. The Sudd has a major impact on the hydrology of the White Nile in two ways. First, the wetland evapotranspires approximately 50 per cent of the runoff and second, the wetlands act as a damper and releases the flow evenly over the years as seen in Figure 8.2. The White Nile contributes approximately 25 per cent of the annual flow of the Nile entering Egypt.

The Blue Nile and Atbara systems have their sources on the Ethiopian highlands and have an entirely different hydrology. They receive the heavy rains of the summer monsoon and produce the 3 month "summer Nile Flood" that reaches Egypt in late July and recedes in late October. The flow of the Blue and Atbara Rivers contribute approximately 75 per cent of the annual flow of the Nile entering Egypt, almost entirely during the summer flood as seen in Figure 8.2.

Paleohydrology

Archeological, geochemical, and palynologial evidence suggests that the Nile River basin and resulting flows have varied greatly due to global and regional climatic variations. From the beginning of the Egyptian Dynastic period, approximately 3500 BC, the Nile basin has remained stable in its current Equatorial and Ethiopian Highland sources. However, there have been major wet and dry periods over the past 5000 years with the current millennium appearing to be the most arid of the late Quaternary. Since the Dynastic period there have been short-term sequences of droughts and floods. These periods have greatly affected the economic and political success of individual dynasties as seen by 14-year periods described in Genesis.

Hydrology of the Last Millennium

The Massachusetts Institute of Technology's Elfatih Eltahir and Guiling Wang created a record of water levels from the Nile River extending back to 622 A.D. and examined the correlation of these flows to large-scale

Figure 8.2

Contribution of Major Nile Tributaries to Flow Entering Egypt

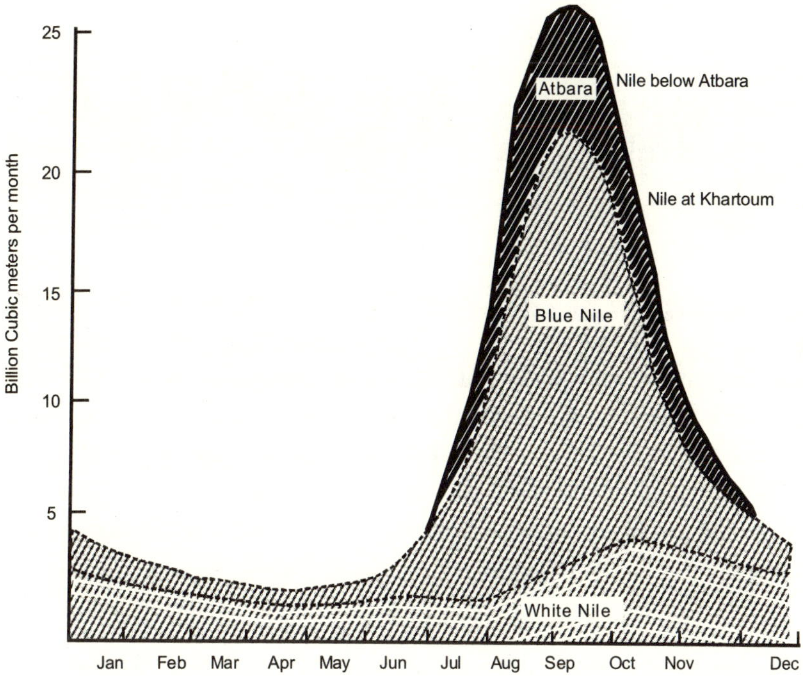

phenomena such as jet stream circulation patterns or the El Niño frequency. The researchers found a strong relationship between the Nile's flow level and the existence of El Niño—so strong that the Nile water level record can be used as a proxy for El Niño occurrence. They found that the frequency of El Niños in the 1990s has been greatly exceeded in the past. In fact, the period from about 700 to 1000 A.D. was a very active time for El Niños, making our most recent decade or two of high El Niño frequency look as mild in its comparison. A graph of the Nile minimum flows from 622 to 1400 shows short-term sequences of droughts and floods that persisted in this millennium (Figure 8.3).

Modern Hydrology

The modern era of Nile hydrology began around 1870 with the introduction of an accurate stream flow gauge at Aswan. Figure 8.4

Figure 8.3

Nile River Minimum Water Levels for 622 AD to 1284 AD

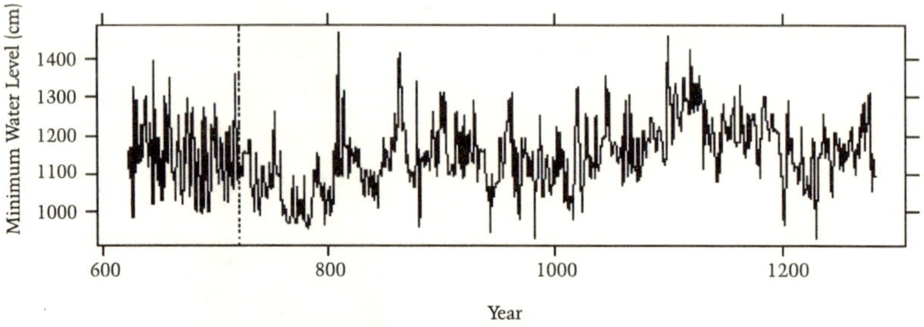

Note: The vertical dotted line is at 730 AD.

shows the stream flow record of the major tributaries of the Nile and other major African rivers for most of the 20th century. It shows that short-term sequences of droughts and floods are seen in the 1900's with a major difference in the mean flow at Aswan from 1870 to 1900 of 110 MCM as compared to 84.9 for 1900 to 1960. It is the record from 1900 that is accepted by hydrologists and water resources planners, of the last

Figure 8.4

Stream Flow of Major Nile Tributaries and Other African Rivers

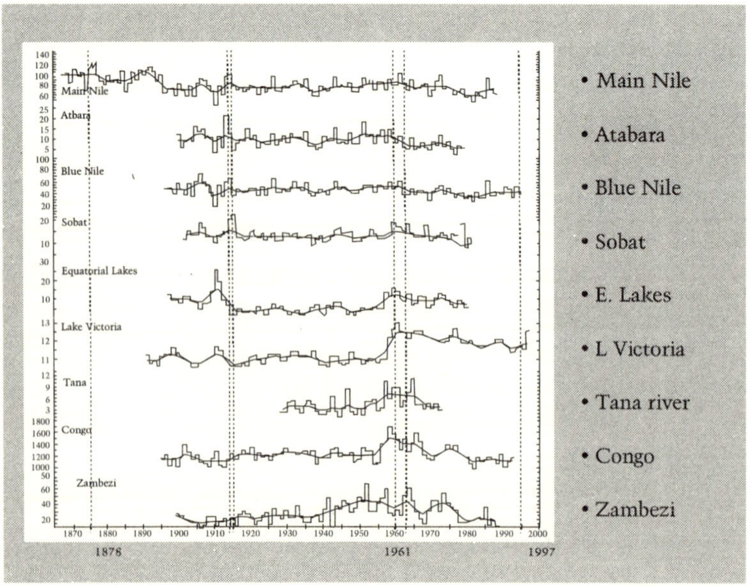

half of the twentieth century, for the development and management of the Nile River resources.

Egypt and the Utilisation of the Nile

The Nile enters Egypt at its southern border and flows almost directly north for 1000 kilometres to Cairo and the start of the Nile Delta and then on to the Mediterranean Sea via a number of river branches (Figure 8.5). The number of river branches has changed over the past 5000 years from seven in the time of Moses to two currently.

Figure 8.5

Map of Egypt

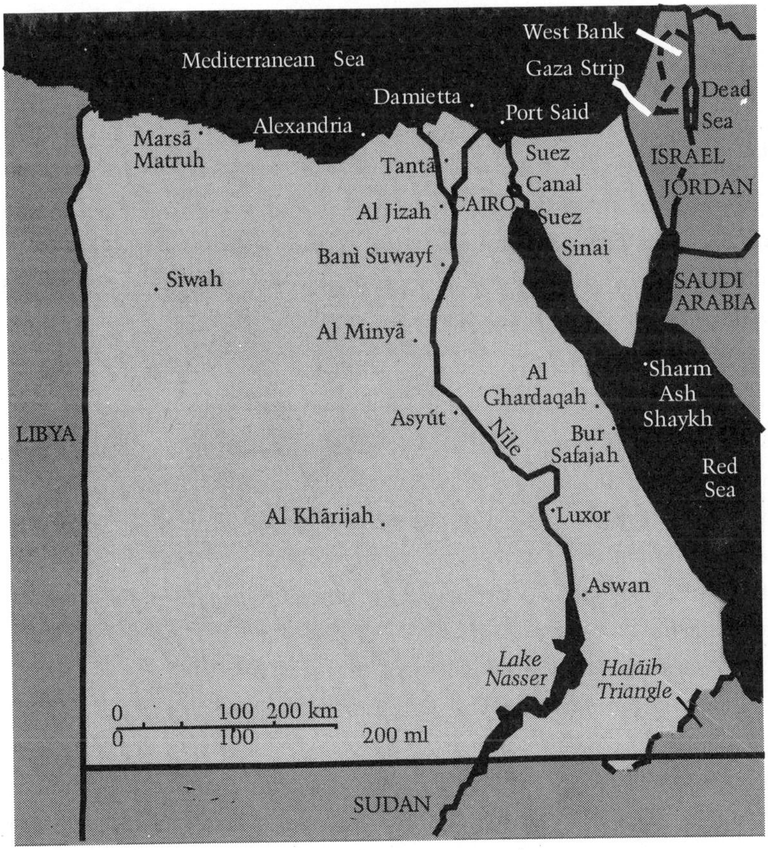

Dynastic

Most of Egypt receives no rainfall and rain fed agriculture is limited to a small strip of land along the Mediterranean coast. Agriculture is limited to the limited land of the Nile Valley that can be irrigated by waters from the Nile. Thus, agriculture is a function of the ability to apply water to the land and the availability of water, which in turn is a function of the type of irrigation system and the management of water resources of the Nile River.

Recession Irrigation

In pre-dynastic time, the Nile flood would overflow its banks starting in August and inundate the lands adjacent to the Nile, saturating the soil and depositing its rich sediments as a fertile layer of new soils. Farmers would sow seeds on the lands as the flood waters 'receded'. This allowed for only one crop per year and that too only during the winter season after the flood.

Basin Irrigation

The first major irrigation engineering development in Egypt occurred during the first Dynasty around 3100 BC with the intentional flooding and draining of land along the Nile by using dykes to form "Basin" with the water controlled by sluice gates.

"The traditional form of irrigation along the Nile was the use of basins, varying in size from 1000 to 4000 feddans. In turn these were divided into smaller units of 4 to 5 feddans running in parallel to the main irrigation canals. The flood wave along the Nile used to begin to affect Upper Egypt by late July. Water levels rose rapidly in August and peaked in September. The water was kept in the basins for 40-60 days and then drained back to the river, or, in times of water shortage, into the next basin downstream" (Hurst, 1952). Once the water had left the fields, no more water was added until the next flood. Therefore, all crop growth was dependant on the soil moisture reserves that had been replenished. With the long dry season the soil was broken up by desiccation and little plowing was necessary before the seeds were sown on the newly deposited silts and clay once the waters had receded. Plant growth took place in the winter period and the crop was harvested in late spring. The ground then remained fallow until the next crop was sown flowing the succeeding annual flood. The traditional basin irrigation system limited the range of crops that could be grown with cereals, by far the most important food crops. The basin irrigation system was designed to make use only of the flood wave. No control of the flow of the river was ever attempted and as a result most of the flood waters were discharged into the Mediterranean, unused.

Figure 8.6 illustrates just how important the Nile river waters are to the Egyptian agriculture. This remote-sensed image demonstrates that vegetation can only occur where the Nile waters have been applied to lands via irrigation.

Figure 8.6

Remote Sensed Image of Nile Delta, October 1986

Lift Irrigation

The second major irrigation engineering development, lifting Nile water via the *shadoof* (human-powered counter weighed bucket on a long pole) occurred during the 18th Dynasty (1550–1307 BC). This concept was enhanced by the introduction of the *saqia,* Persian water wheel around 300 BC. The *saqia* is an animal powered device that could lift significant amounts of water. This technology allowed for supplementing flood flows during drought years and providing for limited summer irrigations.

Summer cultivation was also practiced using waters from flood plain wells or from the Nile itself. The main difficulty with this type of operation was that the water often had to be lifted considerable distances, severely restricting the area that could be cultivated. It did, however, enable valuable crops such as sugarcane, rice, cotton, and tobacco to be grown. This system of irrigation remained almost unchanged until the middle of the 19th century.

The Nineteenth Century

Perennial Irrigation

The first major change that occurred in the irrigation system was the introduction of perennial irrigation. In 1805, the new ruler of Egypt, Mohammed Ali Pasha felt that the future of Egyptian development should be via development of summer agriculture, especially the cash crop, cotton plus sugar cane and vegetables. To achieve this, engineers came up with a plan to build structures (called barrages) across the Nile to raise the water levels of the low summer flows to provide for gravity flow in one-metre deep irrigation canals.

Two barrages were constructed north of Cairo at the start of the Nile Delta on the two branches of the Nile, the Rosetta and Damietta, and are known as Mohammed Ali Barrages. Although there were some problems with the structural integrity of the barrages, the system was very successful and cotton became a major export commodity and led to the beginning of the Egyptian textile industry.

Still, by the end of the nineteenth century, the agricultural system was limited to summer flows that could be extremely low and variable from year to year (Figure 8.7).

Figure 8.7

Mean Monthly Nile Flow at Aswan

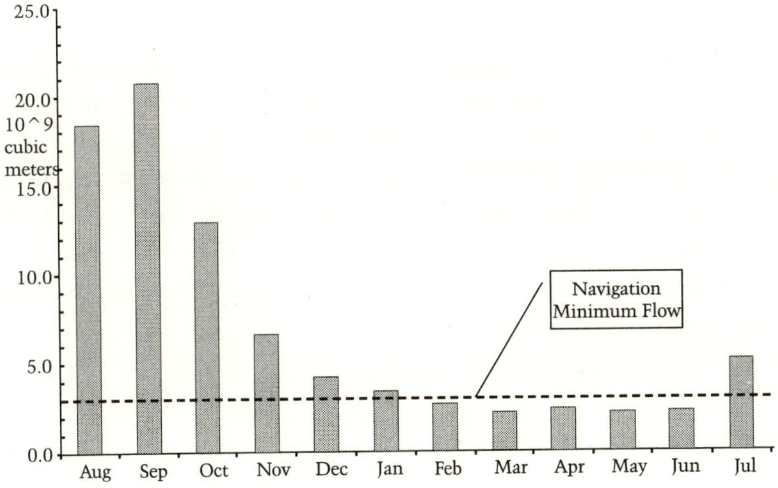

The Twentieth Century

The Aswan Dam

The desire to have more water for the profitable summer crops required the storage of some of the flood waters for release during the summer irrigation season. To achieve this, in 1902 a dam was built at Aswan that held 1 million cubic metres (MCM), equal to about 2 per cent of the average annual Nile flow. It was so economically successful that it was raised in 1912 and then again in 1934 to total storage of 5 MCM. In addition, Egypt obtained an extra 1 MCM of storage in the Jebel el Aulia Dam on the White Nile at Khartoum, Sudan. Egyptian water resource engineers and policy makers dreamt of controlling the Nile flood to avoid its damage and store its waters for the profitable summer crops. This dream became a reality with the completion of a new dam just upstream of the Aswan dam.

The High Aswan Dam

After the July 1952 Revolution, studies were carried out to construct the High Dam at Aswan. The World Bank agreed to finance the huge project but later withdrew its consent. This led Egypt to nationalise the Suez

Canal to use its revenue complemented by technical assistance from the Soviet Union to finance the project.

The Engineering

On January 9, 1960 the foundation stone of the High Dam was laid. On January 15, 1971 a ceremony was held marking the completion of the High Dam. The total cost of its construction was LE 500 million. Fifty thousand engineers and workers took part in the implementation of this giant project. The High Dam is a rock-filled dam south of Aswan. It is 3600-metre long and pyramidal in shape. It is 980 metre wide at the base and 40 metres at the top. It is 111 metre high above the Nile floor and 196 metres above sea level.

Water is released via six huge tunnels, 14 metres in diametre. There are 12 huge units for hydraulic power generators. A great lake has been formed in front of the High Dam, Lake Nasser, with a storage capacity of 164 billion cubic metres that reach 182 metres high. Lake Nasser is a large reservoir in the midst of the desert with a large surface area, which leads to enormous amounts of evaporation.

In 1959, Egypt and Sudan signed the Nile Waters Agreement of 1959, a bilateral agreement to allocate the water of the Nile potentially crossing the Egyptian-Sudanese border. The agreement assumed the following availability and proposed allocations as shown in Table 8.1.

Thus, Egypt is legally able to annually release 55.5 MCM. With the active storage of the High Aswan Dam, reservoirs design theory suggests that a firm yield (a constant annual release with 100% reliability) of 55.5

Table 8.1

Water Allocation under Egyptian-Sudanese Bi-lateral Agreement (in BCM)

Average Annual Nile Flow	**84.0**	
Reservoir Losses Due to Evaporation and Seepage		-10.0
Net Water Availability per Annum		74.0
Allotment to the Sudan		18.5
Allotment to Egypt		55.5
Total Water Usage per Annum		74.0

MCM is obtainable at the cost of evaporating 10 MCM or 12.5 per cent of the annual flow. Figure 8.8 shows the impact of High Aswan Dam on water supply availability to Egypt.

Figure 8.8

Impact of High Aswan Dam on Egyptian Water Supply

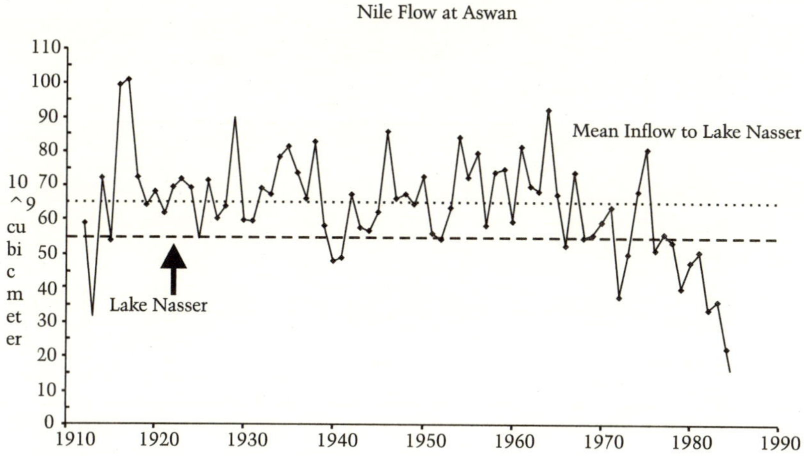

The Benefits

Economic

The economic benefits of providing a highly reliable and non-flooding water supply to Egypt are listed below:

- It has saved Egypt from devastating floods which resulted in lost summer harvests and damage to infrastructure, and potential loss of life.
- The High Dam water has been used in reclaiming about 1.2 million feddans.
- Perennial irrigation of about 850 thousand feddans has replaced basin irrigation.
- Rice and sugarcane production has increased considerably.
- The High Dam turbines generate an average of 8 billion kilowatt-hour used in industry and the electrification of all towns and villages in Egypt.

- The High Dam has facilitated navigation up and down the Nile all the year round.

Table 8.2 shows the actual crop areas in 1995, 25 years after the High Aswan Dam, compared with Pre-Dam area. It shows major increases in wheat, maize, rice and sugar cane areas, and a major decrease in the cotton area.[1] Also, Table 8.3 shows a typical annual cropping calendar after the dam.

Table 8.2

Cropped Area

	1960	*1995*
Wheat	1,387	1829
Maize	1,727	1906
Millet	469	346
Rice	799	1276
Cotton	1,751	884
Sugar Cane	122	274
Total	6,255	6,515

Table 8.3

Cropping Calendar of Egyptian Major Crops

	Aug	*Sep*	*Oct*	*Nov*	*Dec*	*Jan*	*Feb*	*Mar*	*Apr*	*May*	*Jun*	*Jul*
Berseem			1	1	1	1	1	1	1	1		
Short-berseem			1	1	1	1	1	1				
Wheat				1	1	1	1	1	1	1		
Barley				1	1	1	1	1	1	1		
Winter veg			1	1	1	1	1	1	1			
Cotton	1	1	1					1	1	1	1	1
Rice	1	1	1							1	1	1
Summer maize	1	1								1	1	1
Sorghum	1	1								1	1	1
Summer veg	1							1	1	1	1	1
Nili maize	1	1	1	1								1
Nili veg	1	1	1	1								1
Citrus	1	1	1	1	1	1	1	1	1	1	1	1
Sugarcane	1	1	1	1	1	1	1	1	1	1	1	1

1. The net addition in cropped area before and after the dam is less than expanded irrigation figures mentioned above. This is mainly attributed to urban encroachment on cultivated land and the difficulties encountered with reclaiming new lands.

Social

It is difficult to measure the social benefits of the High Aswan Dam. However, summarising the proceedings of a one-day symposium held in Cairo in 1993 on the pros and cons of the High Aswan Dam, the then Secretary-General of the International Commission of Large Dams, Joannes Cotillon, made some telling remarks:

- That the stored water saved Egypt from famine in 1972-73 and again in the nine successive drought years from 1979-1987.
- That the reservoir protected the Nile valley from major floods in 1964, 1975 and 1988.
- That it enabled double the previous population to be fed, avoiding the social cost of drought and floods as well as the positive benefits of providing adequate nutrition to the population

Secondary Impacts of the Dam

Fertiliser Impact of Flood and Sediments

It is argued that the dam created demand for fertiliser use as it blocked the sediments that came with the floodwaters and prevented agricultural land from its benefits. However, increased use of fertilisers within Egyptian agriculture was to happen with or without the dam. A number of references state that the fertiliser effect of the Nile flood is a long-term process, and that about 70 to 80 per cent of the sediments were washed into the Mediterranean Sea. The soil-enhancing process of the Nile is of a 100-year or greater time scale. In addition, animal and industrial fertiliser use were greatly increasing even before the High Dam was built. The reasons for such a rapid growth of commercial fertiliser after the Dam came into being can also be attributed to the following:

1) The co-incidence of Green Revolution varieties that required high chemical inputs.
2) The control of the Nile flood prevented the flooding of summer crops and greatly increased the expected value of fertiliser applications.

Impact of High Floods on Agriculture Output

After extensive research with many secondary references we have come up with a robust initial estimate of a Flood Damage Function for Agricultural Output.

Positive impacts of high flood: There is a threshold amount of flood water that can be economically utilised and high floods are just surplus resources with zero to negative value.

One expects that a high volume flood would flood more land and is a potential for more water to infiltrate into soil moisture. However, because of the rare nature of the high flood, there is a limited amount of land served by canals and irrigation due to expected value of the low frequency of irrigation actually taking place, so there is a maximum amount of land that becomes a constraint to the amount of water that can be utilised. Also, to infiltrate more water means leaving the land flooded for a longer period. However, the longer the period the land is flooded, the shorter is the growing period, so there is a trade off in maximising yield between available soil moisture and the length of the growing season.

Negative impacts of floods: For agricultural outputs there are three negative effects, two of which we can quantify for this analysis.

1) Damage to agricultural infrastructure (fixed capital, canal, bridges, pumping stations...), which will not be modeled.
2) A high flood highly correlated to an early flood, leading to a premature harvesting of the summer crops, reducing yield—cotton and maize being the main victims.
3) An extremely high flood can result in a total flooding of the summer crops, while still in the field, leading to a loss of the entire crop.

We have developed a crop damage function from empirical data to model the lost harvest effect of the high floods for summer crops only. The functions is:

$$\left.\begin{array}{l} Y_t = 0 \\ Y_t = \left(\dfrac{Q(f)-68}{13}\right)^{1.5} \\ Y_t = 1 \end{array}\right\} \begin{cases} Q(f) < 68 \text{ billion cubic meter} \\ 68 < Q(f) < 81 \\ Q(f) > 81 \end{cases}$$

Where YRt, = yield reduction for t = summer

and

Q(f) = quantity of flood

Impact on Navigation and Transportation

The construction of the dam has positively affected navigation along the Nile in two ways. First, it is now possible to maintain stable water levels above a minimum, which allows year round navigation from the Mediterranean in the north to Aswan in the south, and lowers navigation fuel costs when the flow is high during the flood season. Second, navigation south of the dam, from Aswan to Dongola in Sudan, is possible through Lake Nasser.

Navigation is a binary process. When flows are below the draft depth no shipping is possible. In terms of modeling the effect of improved navigation can be represented as:

If *Q(f)* = *quantity of flood*

then no shipping when *Q(summer)*< 18 BCM

In 1997, transportation along the Nile River accounted for 8.4 per cent of all shipping. As with hydropower, this shipping can be made up by the higher cost of rail or road. Thus if:

no shipping when *Q(summer)* < 18 BCM then 4.2 per cent are loses to transporation

Impact on Tourism

The High Aswan Dam, allowing year round navigation up the Nile, has led to major investment in Nile River Cruises as well as Lake Nasser Cruises. In 2000, tourism accounted for 4.4 per cent of GDP. A rough estimate is that 25 per cent of Tourism is Nile Related. Hence, as *Q(summer)* falls below 18 BCM, then navigation is disrupted for half of the year. This leads to:

12 per cent loss to tourism when $Q(summer) < 18$ BCM

Also in very high flood years, tourist sites are flooded (especially places in Luxor and the Valley of the Kings). In this case:

12 per cent loss to tourism when $Q(summer) > 81$ BCM

Overview of CGE model Structure and Database

Model Structure

CGE models are solvable numerically and provide a full account of production, consumption and trade in the modeled economy. Since the first applications in the mid-1970s, this class of models has been widely used in policy analysis in developing countries. The present analysis is based on an extended version of IFPRI's Standard CGE model, written in the GAMS (General Algebraic Modeling System) software. The computer code separates the model from the database—with a social accounting matrix (SAM) as its main component—making it easy to apply the model in new settings.

The Standard model follows the disaggregation of a SAM and explains all payments that are recorded in the SAM. It is written as a set of simultaneous equations, many of which are non-linear. There is no objective function. The equations define the behaviour of the different actors. In part, this behaviour follows simple rules captured by fixed coefficients (for example, *ad valorem* tax rates). For production and consumption decisions, behaviour is captured by non-linear, first-order optimality conditions of profit and utility maximisation. The equations also include a set of constraints that have to be satisfied by the system as a whole but which are not necessarily considered by any individual actor. These "system constraints" define equilibrium in markets for factors and commodities, and macroeconomic aggregates (balances for savings-investment, the government, and the current-account of the rest of the world).

The standard model is characterised by flexible disaggregation and pre-programmed alternative rules for clearing factor markets and macro accounts. Figure 8.9 provides a simplified picture of the links between the major building blocks of the model. The disaggregation of activities,

Figure 8.9

Structure of Payment Flows in the Standard CGE Model

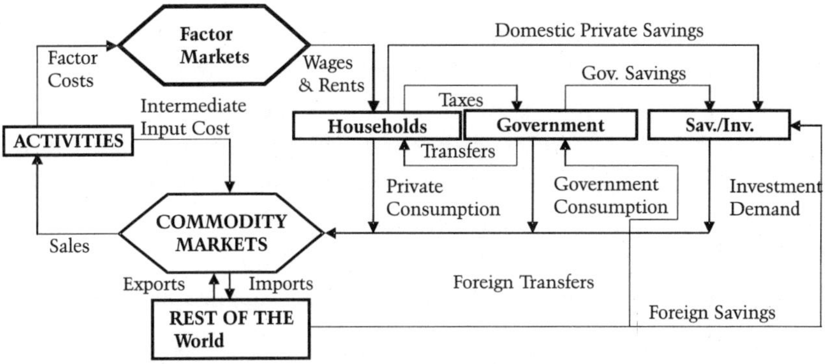

(representative) households, factors, and commodities—the blocks on the left side of the figure—is determined by the disaggregation of the SAM. The arrows represent payment flows. With the exception of taxes, transfers and savings, the model also includes "real" flows (a factor service or a commodity) that goes in the opposite direction. The activities (which carry out production) allocate their income, earned from output sales, to intermediate inputs and factors.

The producers are assumed to maximise profits subject to prices and a nested technology in three levels. At the top of the nest, output is a Leontief or CES function of aggregates of value-added and intermediate inputs. At the second level, aggregate value-added is a constant elasticity of substitution (CES) function of factors, whereas the aggregate intermediate input is a Leontief function of disaggregated intermediate inputs. A third level is specified in one area: agricultural land and water are combined in fixed proportions to form a land-water aggregate that enters as a factor on the second level. Each activity produces one or more commodities and any commodity may be produced by more than one activity.

Given the assumption that they are small relative to the market, producers take prices as given when making their decisions. After meeting home consumption demands, the outputs are allocated between the domestic market and exports in shares that respond to changes in the ratio between the prices that the producers receive when selling domestically and abroad. In the world markets, the supplies of exports are absorbed by infinitely elastic demands at fixed prices (the small-country assumption). Domestic market demands (for investment, private consumption, government consumption, and intermediate input use) are met by supplies from domestic producers and the rest of the world (imports). For any commodity, the ratio between the demands for imports and domestic output responds to changes in the relative prices of imports and domestic output that is sold at home. In world markets, import demand is met by an infinitely elastic supply of imports at fixed prices. In the domestic markets for products of domestic origin, flexible prices assure that the quantities demanded and supplied are equal.[2]

The factor costs of the producers are passed on as receipts to the household block in shares that reflect endowments. In addition to factor incomes, the household block may receive transfers from the government (which are CPI-indexed), the rest of the world (fixed in foreign currency), and from other households. These incomes may be spent on savings, direct taxes, transfers to other institutions, and, for the RHs, consumption. Savings, direct taxes, and transfers are modeled as fixed income shares. Consumption is split across different commodities, both home-consumed and market-purchased, according to LES (Linear-Expenditure-System) demand functions (derived from utility maximisation).

The government receives direct taxes from the households and transfers from the rest of the world (fixed in foreign currency). It then spends this income on consumption (typically fixed in real terms), transfers to households, and savings. The rest of the world (more specifically the current account of the balance of payments) receives foreign currency for the imports of the model country, and then spends these earnings on exports from the model country, transfers to the model country's government, and on "foreign savings" (i.e. the current account deficit). Together the government, enterprises, and the rest of

2. In terms of functional forms, the Standard model uses a CES function to capture the aggregation of imports and output sold domestically to a composite commodity, and a CET function to capture the transformation of output into exports and domestic sales. Without any change in the GAMS code, the model can handle databases with commodities that are only exported (no domestic sales of output), only sold domestically (no exports), or only imported (no domestic production).

the world may play an important role in the distributional process, by "filtering" factor incomes on their way to the RHs and by directly taxing or transferring resources to the RHs. Finally, the savings-investment account collects savings from all institutions and uses these to finance domestic investment.

The user has the option to choose among a relatively large number of pre-programmed alternative closure rules for the factor markets and the three macro accounts of the model, the (current) government balance, the balance of the rest of the world (the current account of the balance of payments, which includes the trade balance), and the savings-investment balance. The closure rules that are selected in this study are presented in a later Section below.

The model is used for comparative static analysis, implying that the impact of the shock (or the combination of shocks) that is being simulated is found by comparing the model solutions with and without the shock(s). Each model solution provides an extensive set of economic indicators, including GDP; sectoral production and trade volumes; factor employment; consumption and incomes for representative households; commodity prices; and factor wages.

Data

Social Accounting Matrix (SAM) for 1996-97

The bulk of the model data are arranged in the format of a disaggregated SAM (an 83x83 matrix) for 1996/97—see Box 8.1. The SAM, which was used in Lofgren and El-Said (2001), was constructed on the basis of data from various official publications including national accounts, government budget, and trade data as well as Egypt's most recent official Social Accounting Matrix (Central Bank, 1995 and 1998, CAPMAS, 1996a, 1996b, and 1998, IMF, 1998). The Egypt Integrated Household Survey (EIHS) and IFPRI research documents based on the EIHS were the primary source for data on consumption (IFPRI, 1998). Data in Kherallah *et al.* (1998) were used for flour production. Underlying the construction of such a SAM is an attempt to make the best possible use of available scattered data. Inevitably imbalances appear when data from different sources and years are integrated in one framework; a SAM-

Box 8.1

Social Accounting Matrices

A social accounting matrix (SAM) provides much of the data needed to implement a computable general equilibrium (CGE) model. A SAM is a square matrix that, for a period of time (typically one year), accounts for the economy-wide circular flow of incomes and payments. It summarises the structure of an economy, its internal and external links, and the roles of different actors and sectors. Its disaggregation is flexible and may depend on data availability and the purposes for which the SAM will be used. If the SAM is to support analyses of Dam projects, it must include a detailed disaggregation of agricultural factors, activities and commodities as well as other areas, such as electricity production, that are most directly affected by the Dam. A SAM brings disparate data (including input-output tables, household surveys, producer surveys, trade statistics, national accounts data, balance of payments statistics, and government budget information) into a unified framework. In order to overcome data inconsistencies, IFPRI has extended estimation methods in 'maximum entropy econometrics' (appropriate in data scarce contexts) and applied them to SAM estimation (Robinson *et al.*, 2001).

Entropy programme, developed at IFPRI, was used to generate a balanced model SAM that retains as much as possible of the information contained in the original data set (Robinson, Cattaneo and El-Said, 2001; Thissen and Lofgren, 1998).

For each of the ten households, income and price elasticities for disaggregated foodstuffs and aggregated nonfood consumption have been taken from Yohannes and Bouis (1999). A variety of sources were used for other elasticity estimates needed for the household nonfood LES functions as well as the functions for import aggregation, domestic output transformation (CET), production (CES), and (constant-elasticity) export-demand.

Compared to the original SAM, we introduced the following changes, suppressing aspects of Egypt's economy that are peripheral to the purposes of this analysis:

- food subsidies have been turned into transfers to the households;
- aggregation of activities and commodities that were disaggregated on the basis of whether they involve or do not involve subsidised commodities.

Table 8.4 shows the SAM disaggregation of institutions, factors, and activities in the model. Among the factors, labour and capital are used by all sectors, while water, summer land, and winter land are used only by agricultural crop activities. The crop activities are differentiated according to the period of land occupation into winter crops, summer crops, and perennial crops. Outside of agriculture, there is a one-to-one mapping between activities (the producing sectors) and commodities (the outputs produced). Inside agriculture, the two berseem activities and the two vegetable activities are both assumed to produce the same commodity (berseem and vegetables, respectively). Given the quality difference between domestic maize (some 95 per cent white maize) and imported (yellow) maize, the latter is a separate imported commodity without any domestic production. Moreover, several crop activities produce byproducts that are used as animal feed. This disaggregation of agriculture makes it possible to capture direct links between crop and animal activities: crop outputs (most importantly berseem, maize, and various crop by-products) are used as inputs in the animal activity; animal outputs (manure and animal labour) are used as inputs in crop activities. Table 8.5 shows an aggregate version of the 1997 Egypt SAM, and Table 8.6 reports selected shares showing the structure of the Egyptian economy in the SAM.

Table 8.4

Disaggregation of Factors, Institutions and Activities

Set	*Elements*	
Institutions (12)	Households Government Rest of the world	rural and urban, both disaggregated by quintile
Factors of production (5)	Capital Labour Water Summer land Winter land	agricultural and nonagricultural agricultural and nonagricultural
Activities (26)	Winter crops	wheat, long berseem, short berseem, legumes, winter vegetables, other winter crops
	Summer crops	cotton, rice, maize [including sorghum], summer vegetables, other summer crops
	Perennial crops	**fruits, sugarcane**
	Other agriculture and food processing	animal agriculture, bread, flour, other food processing
	Other	oil, cotton ginning, textiles, other industry, electricity, construction, government services, **transportation, other services**

Table 8.5

Egypt Macro SAM, 1997 (Billion L.E)

	- 1 -	- 2 -	- 3 -	- 4 -	- 5 -	- 6 -	- 7 -	- 8 -	- 9 -	- 10 -	- 11 -	*Total*
1. *Activity*		437.54										437.54
2. Commodity	196.30			196.25	26.01	54.58	45.08					518.22
3. Factors	238.81											238.81
4. Households			231.44		13.65	11.21						256.30
5. Government			11.34			-0.08		14.55	8.17	2.43	10.55	46.96
6. Rest of the World		61.96	-3.98									57.99
7. Saving-Investment				45.50	7.30	-7.72						45.08
8. Institution tax				14.55								14.55
9. Import tax		8.17										8.17
10. Activity tax	2.43											2.43
11. Commodity tax		10.55										10.55
Total	437.54	518.22	238.81	256.30	46.96	57.99	45.08	14.55	8.17	2.43	10.55	

Source: Lofgren and El-Said (2001).

Table 8.6

Structure of the Economy, Egypt 1999

Sector	*Value added*	*Output*	*Employment*	*Exports*	*Export/ Output*	*Imports*	*Import/ final demand*
Agriculture						(%)	
Wheat	1.74	1.19	1.39	–	–	5.97	42.28
Maize	1.56	1.02	1.35	–	–		
Yellow maize	–	–	–	–	–	2.10	100.00
Rice	1.05	0.76	0.89	0.13	2.10	–	–
Legumes	0.20	0.17	0.24	0.03	2.57	0.31	30.98
Sugarcane	0.37	0.29	0.40	–	–	–	–
Other winter crops	1.55	1.12	2.02	–	–	–	–
Other crops horticulture	1.10	0.80	1.47	–	–	–	–
Fruits	2.01	1.42	1.59	0.29	2.51	0.20	2.73
Vegetables	3.10	2.34	1.75	0.17	0.92		
Cotton	1.09	0.74	1.53	–	–	–	–
Berseem	0.89	0.54	0.23	–	–	–	–
Livestock	3.07	4.00	1.87	–	–	1.08	4.72
Industry							
Bread	0.21	1.20	0.49	–	–	–	–
Flour	0.30	1.53	0.17	–	–	0.12	1.57
Other food	2.66	9.02	1.67	1.81	2.50	3.14	6.09
Oil	6.61	4.38	0.87	18.33	52.26	9.19	38.74
Cotton ginning	0.04	0.91	0.08	0.58	8.00		
Textiles	5.64	6.93	7.53	6.68	12.02	1.88	4.87
Other industry	9.31	12.91	6.19	10.34	10.00	62.85	47.25
Electricity	1.78	1.62	1.48	–	–	–	–
Construction	5.34	7.70	5.19	–	–	–	–
Services							
Government services	7.85	5.83	24.15	–	–	–	–
Transportation	6.78	6.49	4.12	22.06	42.41	1.58	5.65
Trade	35.76	27.12	33.34	39.57	18.20	11.57	6.88
Other services	7.85	5.83	24.15	–	–	–	–
Total	100.00	100.00	100.00	100.00	12.47	100.00	15.48
Agriculture	17.72	14.36	14.73	0.62	0.54	9.66	9.69
Non-agriculture	82.28	85.64	85.27	99.38	14.48	90.34	16.52

Source: Egypt 1997 SAM.

Other Data

Additional data used in the different simulations include:

Nile Flows

To match the cropping season of Egypt as well as the seasonality of the Nile flows, we have divided the Nile into two seasonal flows, Flood from August to January and Summer from February to July. The flood flows account for approximately 70 per cent of Nile flow at Aswan and primarily come in the three months of August, September, and October with its source the Blue Nile with headwater in Ethiopia. Summer flow comes almost exclusively from the White Nile, which exits the SUDD wetlands in Southern Sudan after losing 50 per cent of its volume to evapo-transpiration by wetland vegetation.

With the Nile Waters Treaty, Sudan is entitled to abstract 18.5 BCM, but has averaged only 12 BCM per year since 1965. Without the High Aswan Dam, Egypt can store a total of 6 BCM of flood flow for use during the summer season. The correlation coefficient between the two seasonal flows is only 0.32, which is expected as the meteorological process driving the flows, are spatial and temporally different. Table 8.7 gives the historical time series of seasonal natural flows and seasonally adjusted flows to reflect Sudanese abstractions and within-year storage at the Aswan dam (4 BCM) and Gabel el Ali Dam (2 BCM) in Sudan.

Table 8.7

Nile Flow

	Natural Nile Flow at ASWAN with no abstraction by Sudan or storage			*Available Nile Flow at ASWAN after 12 BCM abstraction by Sudan and storing 6BCM from Flood in Aswan and Gabel Reservoirs*		
	Flood	*Summer*	*Total*	*Flood*	*Summer*	*Total*
Mean	67.42	16.96	84.37	55.42	16.96	72.37
Standard Deviation	13.18	3.48		13.18	3.48	
Covariance	0.20	0.21		0.24	0.21	

Water Use in Egypt

Shiklamanov (2000) reports the distribution of Egyptian Water use in 1995 as listed in Table 8.8 below. As seen in the table, municipal and industrial (M&I) water use accounts for 15.3 per cent of withdrawal and 2.7 per cent of consumption. Government policy has been and remains such that these uses get priority over agricultural uses. These uses tend to be evenly distributed over the year. Thus, for the purpose of water balance modeling it is suggested that the following seasonal M&I water demands be assumed:

Table 8.8

Water Use: 1995

	Withdrawal		*Consumption*	
		BCM/year		
Water withdrawal:	55.5		37.6	
Distribution over sectors:		% BCM/year		
Agricultural	47.0	84.7	30.6	96.7
Domestic	3.6	6.5	0.5	1.5
Industrial	4.9	8.8	0.6	1.8

Sectors: (BCM)	*Flood (BCM)*	*Summer*
Municipal	1.8	1.8
Industrial	2.45	2.45

Table 8.4 indicates that the SAM has 12 agricultural sectors grown over three seasons. Winter crops grown traditionally in the lower temperature months of October to March, and before the High Dam, were irrigated by basin irrigation from the floodwaters. These crops include: Wheat, Long Berseem, Short Berseem, Legumes, Winter Vegetables, and Other winter plants.

The summer crops grown from April to September were traditionally grown with water stored from the flood in the Aswan and Gabel dams, plus diverting via deep canal summer flows and then lifting the water on the land to irrigate. These crops include: Cotton, Rice, Maize, Sorghum, Summer Vegetables, and other Summer Plants.

Finally there are the perennial crops, which are citrus orchards and sugar cane. These crops occupy the land the year round. For purpose of water balance modeling the water required for perennial crops has been divided into their winter and summer water demands. Perennial crops require only 35 per cent of their water withdrawal in the winter six months while the warmer summer requires 65 per cent.

Water: Data and Modeling Approach

Modeling Crop Water Use With and Without Dam

A recent award winning paper by Ali and Mahmoud (n.d.) by the Ministry of Public Works and Water Resources, Government of Egypt provides us with the last detailed look at crop water consumption and cropping patterns for 1993. Based on their data and the crop sector disaggregation in the SAM, a calibrated water use sector is presented in Table 8.9. They assume that overall hydraulic efficiency for the Egyptian irrigation system from Aswan to the Mediterranean Sea is 65%. This includes all water reuse along the Nile and in the Delta. Based on that

Table 8.9

Agriculture Water Use: 1995

Crop sectors	*Crop Water Consumption (m^3 per feddan)*	*1993 Crop Area Feddans*	*Water Consumption (BCM)*	*Water Withdrawal (BCM)*
Winter crops				
Wheat	1,816.38	1,829,232	3.3	5.1
Legumes	1,461.37	323,700	0.5	0.7
Long Berseem	2,538.18	1,668,846	4.2	6.5
Short Berseem	903.47	642,643	**0.6**	0.9
Winter vegetables	1,489.10	508,040	0.8	1.2
Other Winter	1,286.65	75,429	0.1	0.1
Perennial-winter	2,030.51	794,032	1.6	2.5
Total Winter		5,841,922	11.1	17.0
Summer crops				
Cotton	3,025.81	884,310	2.7	4.1
Rice	4,691.40	1,276,295	6.0	9.2
Maize+Sorghum	2,525.84	2,252,043	5.7	8.8
Summer Veg	1,939.89	558,549	1.1	1.7
Other Summer	2,473.08	286,106	0.7	1.1
Perennial summer	3,770.94	794,032	3.0	4.6
Total Summer		5,257,303	19.1	29.4
Total annual		11,099,225	30.2	46.5

assumption and the calculated crop consumption we can determine crops' water withdrawal requirements.

In the CGE model, water and land usage are assumed to be in fixed shares to produce crops—a Leontief production function. In addition, we specify four possible land-water technology combinations for each crop, with assumed declining marginal productivity of water, as shown in Figure 8.10. In effect, the model specifies the choice of land-water combinations by crop as a linear programming problem.[3] The farmer chooses the least-cost land-water technology for each crop, given the 'prices' of land and water.[4] The model solves for land rental rates and the shadow price (or scarcity price) of water that ensures the efficient allocation of available water and land across competing uses (i.e., crops in this model, since non-agricultural demand for water is assumed to be met always).

The model distinguishes two types of land and water: summer land, winter land, summer water, and winter water. With the exception of vegetables and sugar cane, the various crops are season-specific—they are either grown in the winter or summer, and only use land and water

Figure 8.10

Land/Water Technology

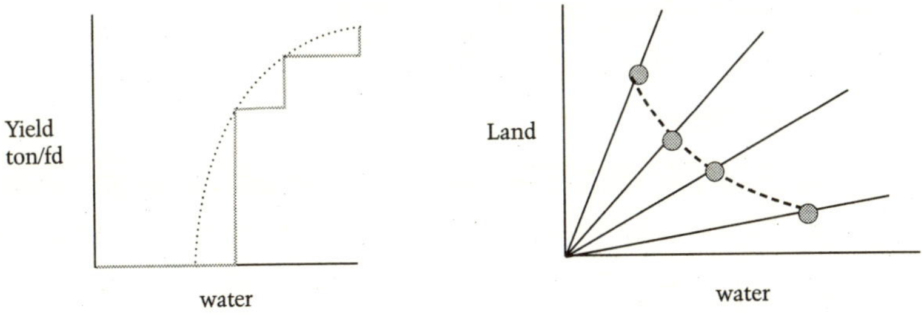

3. Berck, Goldman, and Robinson (1991) and Robinson and Gehlhar (1996) use this technique in CGE water models, but do not assume the availability of alternative technologies by crop.

4. The linear programming problem is translated in the CGE model into a mixed complementarity problem (MCP) and the overall model is solved using an MCP algorithm (PATH), with the linear programming complementary slackness conditions explicitly specified. See Lofgren and Robinson (1999).

in one season. The aggregate supplies of winter and summer land and water are specified separately, and the model generates separate shadow prices for winter and summer land and water. Since the effect of the Aswan dam is to permit Egypt to make the seasonal flows of water more even over the year, the model is specified so that, for the with-dam scenario, the total High Aswan Dam release of 55.5 BCM is available to be allocated to either summer or winter crops. Without the dam, the winter flood is assumed to generate an excess supply of water in the winter, with corresponding water scarcity in the summer.

Given the solution to each sectoral linear programming problem, the model generates a land-water aggregate for each agricultural sector that is then assumed to enter a neoclassical production function along with labour and capital. The rental rate of the land-water aggregate in the model is assumed to consist of separate values of the land and water components, valued at their shadow prices. Agricultural technology in the model is thus specified as a nested function with a Leontief demand for intermediate inputs, a CES function of value added (labour, capital, and land-water aggregate), and a linear programming specification of land and water use. In the base solution, it is assumed that water is in excess supply and hence has a shadow price of zero. In this case, the entire value of the land-water aggregate is attributed to land.

Modeling the Water Balance in the CGE With and Without Dam

With the High Dam

The firm yield of the High Dam is an annual release of 55.5 billion cubic metres. Then we can define the water balance equation as follows:

$$\sum_{s,c} A_c WC_{c,s} + WMI_s \leq TWA$$

$$Q_s = \sum WC_{c,s} + WMI, \leq TWA_s$$

Where s = season c = crop A = area

WC = Leontief water consumption coefficient for the two seasons

TWA = total water available

Q = quantity of water flowing through the turbines

WMI = municipal and industrial water demand (set exogenously)

Since Aswan is downstream and their turbines have less capacity Q_s is appropriate for both power stations (see section on hydropower below). Also for land, the balance equation is defined as:

$$\sum_C A_C \leq Total\ irrigation\ land$$

Without the High Dam

The big difference for water balance is that now we must have seasonal water balances each year and the amount of TWA is a stochastic random variable. The water balance equation is defined as:

$$\sum_C \left(A_C WC_{C,S} + WMI_S \right) \leq TWA_S$$

Turbine flow is directly limited by the season flow and not the irrigation decisions. Plus there is no High Dam power station (see section on hydropower without dam below). In this case the land balance equation is defined as:

$$\sum_C A_C \leq Total\ irrigation\ land$$

Hydropower in the CGE With the High Aswan and Aswan Dams

Table 8.10 below gives a time series of the mix of hydro versus thermal electricity in Egypt from 1980 to 2000, which includes our SAM base year of 1997. Hydroelectric power is the product of flow through the turbines, Q, and, Head (Net difference between water level in the reservoir upstream and water level downstream),

$$E = B^* Q^* H$$

We see lower hydro generation in the mid to late 80's because there was a drought and while the release from High Aswan was maintained the reservoir level was dropping. We see generation going up in the end of the 90's because reservoir levels were going up. We see a dramatic increase in 1999 and 2000 as renovation of the generators completed in 1999 led to increased power capacity.

Table 8.10

Egypt Hydroelectric Power Generation, 1980-2000

	(Billion Kilowatt hours)				
	Hydro	*Thermal*	*Total*	*% Hydro*	*%Low DAM*
1980	9.7	8.56	18.26	53%	11%
1981	10.11	10.39	20.50	49%	10%
1982	10.37	9.57	19.94	52%	10%
1983	10.09	12.39	22.49	45%	9%
1984	9.90	19.42	29.32	34%	7%
1985	8.02	24.28	32.30	25%	5%
1986	7.92	25.56	33.48	24%	5%
1987	8.17	28.58	36.75	22%	4%
1988	7.82	27.58	35.40	22%	4%
1989	7.92	29.78	37.70	21%	4%
1990	9.88	31.53	41.41	24%	5%
1991	8.55	34.02	42.57	20%	4%
1992	8.46	35.04	43.50	19%	4%
1993	10.38	37.37	47.75	22%	4%
1994	10.63	39.43	50.06	21%	4%
1995	10.70	41.38	52.08	21%	4%
1996	11.44	40.32	51.72	22%	4%
1997	11.88	42.96	54.84	22%	4%
1998	12.10	47.11	59.21	20%	4%
1999	15.15	49.54	64.69	23%	5%
2000	15.94	53.65	69.59	23%	5%

The High Aswan Dam power generation function based on flow going through the turbines is given as : if E is annual energy generated in 10^6 kilo watt hour (K.W.H) and Q is flow through the turbines in billion cubic metres (BCM) then as shown in Figure 8.11:

$$E = 201.63^{*} Q - 1.229^{*} Q^2;$$

and the relationship between Q and release from Aswan is as follows:

If $Q_{flood\ season}$ is the flow during the flood season (August–January) and

If Q_{summer} is the flow during the summer season. Then

$Q_{flood\ season} = 27.5$ if $Q_{flood} \geq 27.5$

$$Q_{flood\ season} = Q_{flood\ season} \text{ if } Q_{flood} < 27.5$$

$$Q_{summer} = 27.5 \qquad \text{if } Q_{summer} \geq 27.5$$

$$Q_{summer} = Q_{summer} \qquad \text{if } Q_{summer} < 27.5$$

$$\sum_{s} Q_s = Q_{flood\ season} + Q_{summer} \text{ for s [flood season, summer]}$$

Figure 8.11

Energy Generation from High Aswan Dam

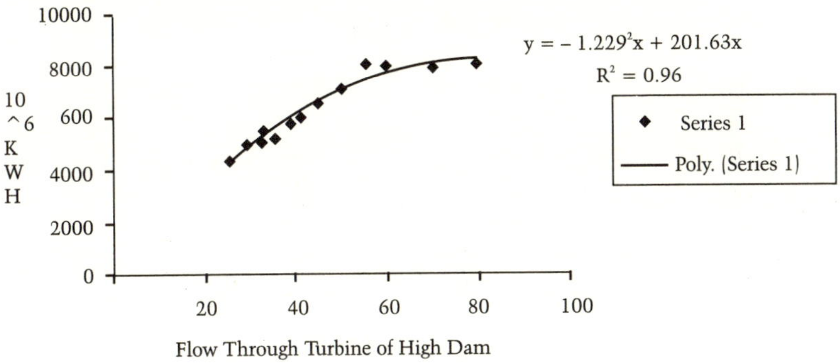

Without the High Aswan Dam but With the Aswan Dam

The Aswan Dam was built in 1902 that held 1 MCM of water which was later raised to store 4 BCM in 1933. It was electrified only in 1960 in conjunction with the building of the High Aswan Dam. It can use more of the Nile flow with a large 182 BCM storage reservoir upstream to smoothen out within and over-year flow variation. The Aswan Dam power station has an installed capacity of 345 MW as compared to 2100 MW of the High Dam power Station. The Aswan turbines have a maximum release capacity of 4 BCM per month (Table 8.11) without the High Aswan Dam. On average, however, more water is spilled than passes through the turbines.

The assumption is that the Aswan Dam would not be removed and thus below is the power generation function based on flow going through the turbines: if E is annual energy generated in 10^6 kilo watt hour (KWh) and Q is flow through the turbines in BCM, then as shown in Figure 8.12:

Table 8.11

Aswan Dam Water Flow and Spill

	Flow	*Turbine*	*Spill*
Jan	3.4	3.4	
Feb	2.7	2.7	
Mar	2.2	2.2	
Apr	2.3	2.3	
May	2.2	2.2	
Jun	2.2	2.2	
Jul	5.1	4.0	1.1
Aug	18.4	4.0	14.4
Sep	20.7	4.0	16.7
Oct	12.9	4.0	8.9
Nov	6.6	4.0	2.6
Dec	4.2	4.0	0.2
		39.0	43.9

Figure 8.12

Energy Generation from Low Aswan Dam: 1933

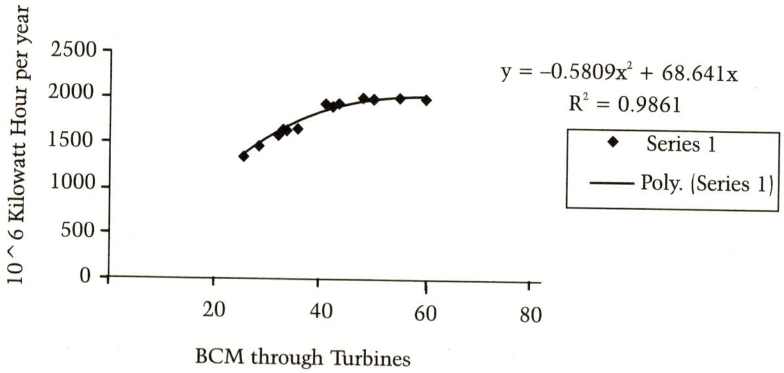

$$E = 68.641 {}^{*}Q - 0.5809 {}^{*} Q^2$$

The relationship between Q and Nile flow at Aswan is as follows:

If $Q_{\text{flood season}}$ is the flow during the flood season (August – January)

and

If Q_{summer} is the flow during the flood season (August – January)

Then

$$Q_{flood\ season} = 24 \qquad \text{if}\, Q_{flood} \geq 24$$

$$Q_{flood\ season} = Q_{flood\ season} \quad \text{if}\, Q_{flood} < 24$$

$$Q_{summer} = 24 \text{if}\, Q_{summer} \geq 24$$

$$Q_{summer} = Q_{summer} \qquad \text{if}\, Q_{summer} < 24$$

$$\sum_s Q_s = Q_{flood\,season} + Q_{summer} \qquad \text{for s [flood season, summer]}$$

Model Simulations

This section presents the results of simulations to assess how Egypt's economy would have performed in 1996/97 without the dam and how the economy, without the dam, would have been affected by year-to-year variations in Nile flows. Variations in Nile flows affect the economy through availability of water supplies for agriculture, the generation of hydropower, navigation, and tourism. Along these lines, we apply a set of historical data about the Nile flows to determine how the Egyptian economy would have fared without the dam.[5] Three sets of simulations are performed using the historical data under different assumptions about factor prices. In all three sets, the 'shock' is to reduce the supply of summer water, using historical data on summer water flows. In effect, the removal of the dam is assumed to force Egypt to adjust to less water in the summer season, with an excess supply of water in the winter.

In simulation 1 (SIM 1), all factor prices adjust to clear factor markets, assuming exogenously specified aggregate employment of labour and capital. In SIM 2, the labour wage is fixed and labour supply adjusts freely to clear the labour market—there is no aggregate employment constraint.

---

5. Appendix A, lists historical data on the Nile flood levels, summer flows, and associated losses in summer agriculture production in the following year, the reductions to navigations from low summer flows, and to tourism from low summer flow. The losses to these activities were determined using the functional relationships described in the previous section. Noting that no dam implied no hydropower, assuming a 25 per cent decline in the electricity sector productivity captured the effect of no dam on hydropower across all historical runs.

In SIM 3, both the wage and the return to capital are fixed; assuming unlimited supplies of labour and capital are available at the fixed wages.

A CGE model with all factor prices fixed, and hence with no supply constraints, operates like a fixed-price, multiplier model. Because factor prices are fixed, then output prices must also be fixed, given standard cost functions. Output is completely demand determined. In fact, this model is not completely demand driven because land and water are assumed to be in fixed supply. The result is a kind of 'constrained multiplier model,' which will behave close to a SAM multiplier model. In SIM 2, with only labour unconstrained, one would expect the multipliers to be smaller than in SIM 3, where both labour and capital are unconstrained. SIM 2 seems like a reasonable specification for a country in which there is excess labour. In SIM 1, all aggregate factor supplies are fixed, and the model will operate like a standard neoclassical CGE model.

In the simulations, the model is run 72 times—one experiment for each specification of winter and summer water availability, based on historical data. Some extreme years were omitted. For these experiments, the output loss in the summer arising from flooding in the previous winter is ignored. These numbers are significant, but we chose to ignore them in experiments since the magnitudes of the crop loss are hard to capture in the model—it is not clear how to model the offsetting market adjustments. So, the experiments we report will underestimate the gains from the dam, or the losses associated with 'removing' the dam.

Tables 8.12 and 8.13 summarise the results for all three sets of simulations. The Base column shows values from the 1997 SAM for Egypt, and the other columns show the mean value and standard deviation of selected variables for the set of simulations using historical flow data. Figures 8.13 and 8.14 show the frequency distribution for real GDP and total absorption for each set of simulations.

Summarising the Results

- Average aggregate economic activity declines under all simulations. The dam generated significant benefits.

Table 8.12

Selected Results for Model Simulations

		SIM1		SIM2		SIM3	
	Base	*Mean*	*Std. deviation*	*Mean*	*Std. deviation*	*Mean*	*Std. deviation*
Real Absorption*	267.3	262.5	4.0	260.1	5.1	233.5	23.5
Consumption	196.3	191.5	4.0	189.0	5.1	162.4	23.5
Exports	54.6	53.5	1.0	53.0	1.1	46.5	5.5
Imports	-62.0	-60.9	1.0	-60.4	1.1	-53.9	5.5
Real GDP	260.0	255.2	4.0	252.7	5.1	226.1	23.5
GDP at factor cost	238.8	235.0	5.2	232.3	4.7	207.9	21.5
Agriculture**	42.3	43.2	4.8	42.0	0.8	39.0	2.9
Winter	12.4	12.5	0.3	12.4	0.1	11.6	0.6
Summer	16.9	17.8	4.8	16.8	0.6	15.8	1.2
Perennial	5.7	5.7	0.1	5.6	0.1	5.1	0.5
Non-agriculture	196.5	191.8	3.0	190.3	4.0	168.9	18.6
Household consumption (L.E. bn.)							
Rural (quintiles):	84.5	82.5	1.9	81.3	2.2	69.0	10.6
1	8.9	8.7	0.3	8.6	0.2	7.4	1.0
2	11.6	11.4	0.4	11.2	0.2	9.7	1.3
3	14.7	14.3	0.4	14.1	0.4	12.0	1.9
4	18.3	17.9	0.4	17.6	0.5	15.0	2.3
5	31.0	30.1	0.6	29.7	0.9	25.0	4.1
Urban (quintiles)	111.8	108.9	2.2	107.8	3.0	93.4	12.9
1	9.1	9.0	0.2	8.9	0.2	7.9	0.9
2	12.9	12.7	0.3	12.6	0.3	11.0	1.4
3	17.5	17.1	0.3	17.0	0.4	14.8	1.9
4	24.0	23.4	0.5	23.2	0.6	20.1	2.8
5	48.2	46.7	1.1	46.2	1.5	39.6	6.0

Notes: *Among the components of absorption, only household consumption changes. Government consumption and investment are fixed in real terms as part of the macro closure rule.

**Includes livestock

- Average real consumption for all households falls across the three simulations.
- The declines increase from SIM 1 to SIM 3. SIM 3, and the full multiplier model generates the largest losses from the removal of the dam. Assumptions about supply constraints affect the results significantly.

Table 8.13

Value Added, Production, Exports and Imports (in L.E. bn.)

		SIM1		SIM 2		SIM3	
	Base	*Mean*	*Std. Deviation*	*Mean*	*Std. Deviation*	*Mean*	*Std. Deviation*
Value added							
Total agriculture	42.3	43.2	4.8	42.0	0.8	39.0	2.9
Winter crops	12.4	12.5	0.3	12.4	0.1	11.6	0.6
Summer crops	16.9	17.8	4.8	16.8	0.6	15.8	1.2
Perennials	5.7	5.7	0.1	5.6	0.1	5.1	0.5
Total non-agriculture	196.5	191.8	3.0	190.3	4.0	168.9	18.6
Total	238.8	266.0	13.0	232.3	4.7	207.9	21.5
Real production							
Total agriculture	62.8	64.0	6.3	62.4	1.1	57.6	4.5
Winter crops	16.2	16.4	0.4	16.3	0.1	15.3	0.8
Summer crops	21.7	23.0	6.4	21.6	0.7	20.4	1.5
Perennials	7.5	7.4	0.2	7.4	0.2	6.7	0.6
Total non-agriculture	374.7	366.3	5.2	363.3	7.3	322.7	35.2
Total	437.5	430.2	18.5	425.6	8.3	380.2	39.6
Exports							
Total agriculture	0.34	0.65	1.48	0.34	0.39	0.50	0.09
Winter crops	0.05	0.07	0.01	0.05	0.07	0.10	0.02
Summer crops	0.14	0.42	1.48	0.14	0.15	0.22	0.05
Perennials	0.16	0.16	0.01	0.16	0.17	0.18	0.01
Total non-agriculture	54.24	52.88	1.17	54.24	52.58	45.99	5.63
Total	54.58	53.53	1.02	54.58	52.97	46.49	5.54
Imports							
Total agriculture	5.99	5.81	0.21	5.72	0.20	4.60	0.94
Winter crops	3.89	3.75	0.15	3.69	0.15	2.86	0.69
Summer crops	1.30	1.28	0.02	1.27	0.03	1.12	0.13
Perennials	0.12	0.12	0.01	0.11	0.00	0.09	0.02
Livestock	0.67	0.66	0.05	0.65	0.02	0.54	0.10
Total non-agriculture	55.98	55.11	0.82	54.63	0.91	49.27	4.60
Total	61.96	60.91	1.02	60.35	1.11	53.87	5.54

Source: Model Simulations

In SIM 1, agriculture gains (especially summer crops with high value) and the burden of the shocks falls on the non-agricultural sectors, with declines in power, transportation, and tourism. The increase in the value of agricultural production arises from the fact that land used in high-value crops has a higher marginal product than land used in low value crops. With the removal of summer water, farmers grow only high value crops, increasing the value of agricultural output. Put another way, the dam allowed Egypt to even the water flow over the year and to support growing low value crops. The dam allowed Egypt to follow a pattern of distorted policies. With uncertain water flows, restricted in the summer, production moves to high value crops, especially summer crops. The decrease in distortions increases efficiency in agriculture, although not enough to offset the losses in electric power, tourism, and transportation. Overall, the dam generates significant gains.

Multiplier models are commonly used in the analysis of the impact of a 'project' such as introducing a dam. The CGE model permits comparison of different kinds of 'constrained' demand multiplier models with a scenario with constraints on factor supplies. For the two multiplier models, SIM2 and SIM3, with less supply constraints:

- the full multiplier model, with both labour and capital unconstrained (SIM 3), yields very large effects.
- the multiplier model with only labour unconstrained (SIM 2) yields much smaller losses than the full multiplier model.

Figures 8.13 and 8.14 show the distribution of results for real GDP and aggregate absorption from the three experiments. Note that the distributions of returns are not symmetric, but are skewed to the left and not unimodal. These results indicate the importance of using data on historical flows in order to determine the pattern of returns from the smoothing of flows due to the dam. Analysing the impact of average changes in flows would miss the wide variation in flows over the historical record.

Using the mean of the results, it is possible to compute the net benefits arising from the dam. Tables 8.14 and 8.15 summarise official Egyptian government analysis of costs and benefits estimated in 1960, and Table 8.16 summarises recent estimates, and includes the estimates of average benefit from SIM 1 and SIM 3. Surprisingly, given the very different approaches taken, the official estimate in 1997 of the annual average net gain (5.5 billion pounds) is within ten per cent of the SIM 1

Figure 8.13

Real GDP Distribution by Simulation

Figure 8.13.1

SIM 1 real GDP distribution
Base = 260.0, Mean = 255.0, Std Dev = 4.0

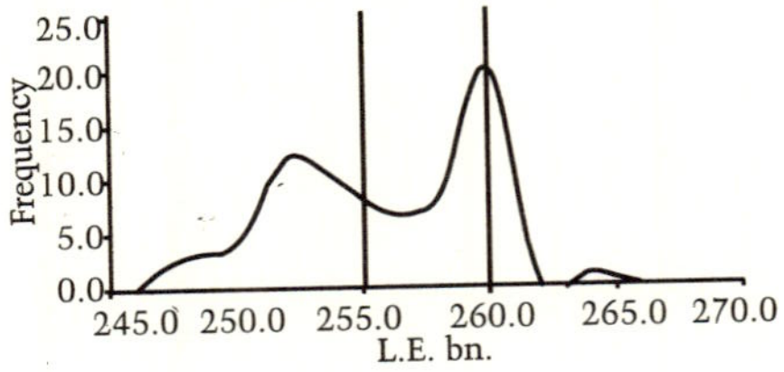

Figure 8.13.2

SIM 2 real GDP distribution
Base = 260.0, Mean = 252.7, Std Dev = 5,1

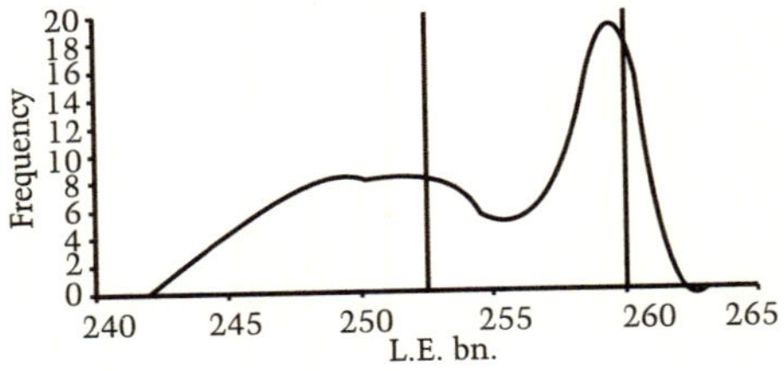

Figure 8.13.3

SIM 3 real GDP distribution
Base = 260.0, Mean = 226.1, Std Dev = 23.5

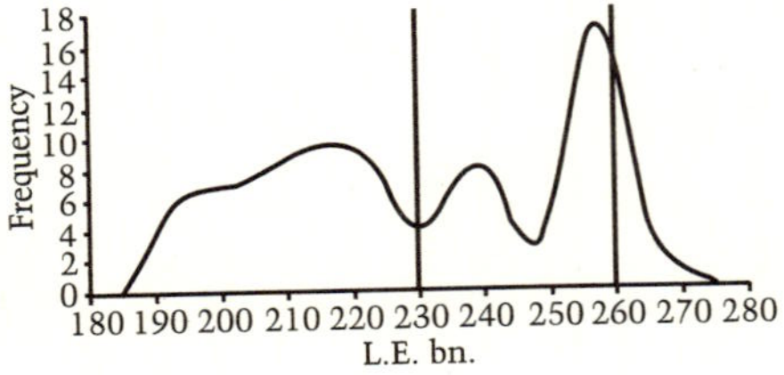

Figure 8.14

Total Absorption Distribution by Simulation

Figure 8.14.1

SIM 1 total absorption distribution
Base = 267.3, Mean = 262.5, Std Dev = 4.0

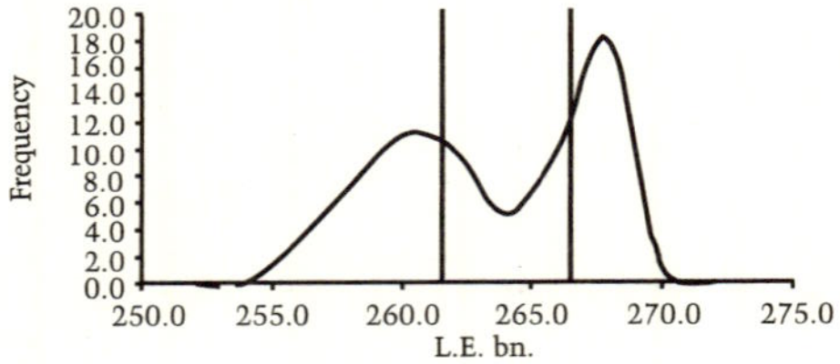

Figure 8.14.2

SIM 2 total absorption distribution
Base = 267.3, Mean = 260.1, Std Dev = 5.1

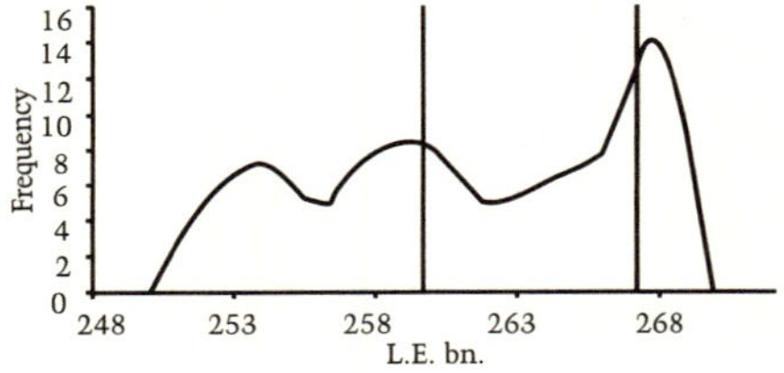

Figure 8.14.3

SIM 3 total absorption distribution
Base = 267.3, Mean = 233.5, Std Dev = 23.5

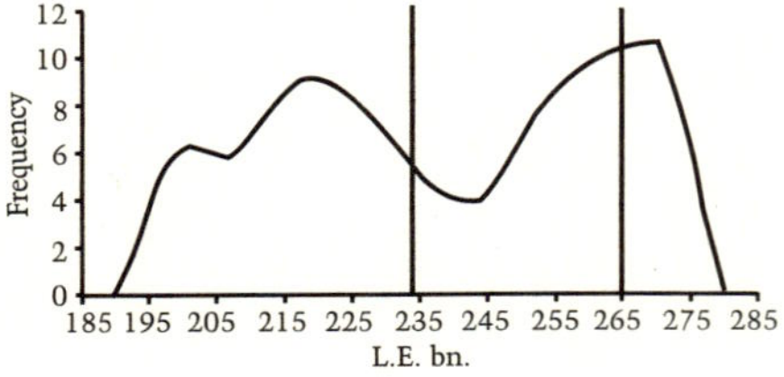

Table 8.14

Estimated Benefits of High Aswan Dam

Annual Increase in GNP (1960 prices)		*L.E. million/year*
1.	Extension of cultivated area by 1,200,000 feddan and conversion of 850,000 feddan from basin to perennial irrigation	63.0
2.	Increasing Rice cultivation to 1,000,000 feddans	56.0
3.	Flood Protection	10.0
4.	Improved Navigation	5.00
5.	Hydropower Electric Energy Generation	100.0
6.	Increase of taxes on Old lands and New taxes	9.0
7.	Savings on barrage maintenance	2.5
8.	Revenue from Dams electrification	10.5
Total Annual Benefit		256.0

Table 8.15

Annual Operating Cost of High Aswan Dam (HAD) Project

Project Element		*Annual Cost L.E. Million (1960)*
1.	Cost of artificial fertilser to replace natural from flood sediment	(a) 20.0 (b) 40.0
2.	Maintenance of Power Station & transmission lines	5.0
3.	Maintenance of HAD	3.0
Total Annual Additional Costs		(c) 28.0 (d) 48.0

Table 8.16

Total Annual Net Benefits

Official Govt. 1960 w/o flood	218.0 L.E. Million (1960)
Official Govt. 1997	5.5 L.E. Billion(1997)
CGE Analysis 1997 w/o flood (SIM 1)	5.0 L.E. Billion(1997)
Multiplier Analysis 1997 w/o flood (SIM 3)	34.0 L.E. Billion(1997)

estimate (5.0 billion pounds). The estimates of net annual gains using the multiplier model, SIM 3, is much higher (34 billion pounds).

Multiplier Results for the Aswan High Dam (AHD)

As shown in Table 8.17 the value-added multiplier values range between 1.22 to 1.4 in the three simulations. The multiplier value of 1.4 implies that for every LE of value-added in directly impacted sectors, another 0.4 LE of value-added is generated through inter-industry linkages and consumption-induced effects.

Table 8.17

Multiplier Results for the Aswan High Dam (AHD)

Value Added (L.E. bn)	*With AHD minus Without AHD*		
	SIM 1	*SIM 2*	*SIM 3*
Value Added of Directly Impacted Sectors*	3.1	5.3	22
Total Value Added	3.8	6.5	30.9
Multiplier	3.8/3.1=1.25	6.5/5.3=1.22	30.9/22.0=1.40

Note: * Directly impacted sectors : agriculture, electricity, navigation and tourism

Income Distribution Impacts of Aswan High Dam (AHD)

In the base case, total consumption is 196.3 LE billion, while under SIM 3 it is 162.4 LE billion, indicating a difference of 33.90 LE billion under a situation 'With Aswan High Dam'. In order to trace the income distribution impacts of the Aswan High Dam, we look at the consumption levels of different quintiles with and without the Aswan High Dam. Figure 8.15 shows quintile-wise consumption of the rural population with and without AHD, and Figure 8.16 shows quintile-wise consumption of the urban population.

Figure 8.15

Rural Consumption Quintile-wise (LE bn) With and Without AHD

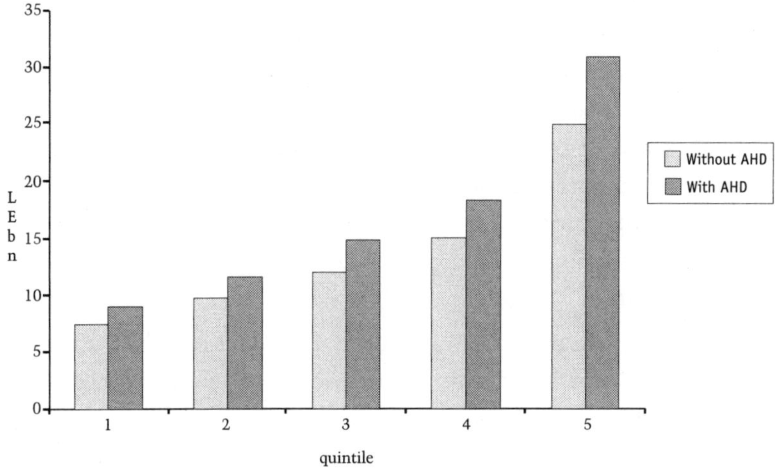

Figure 8.16

Urban Consumption: Quintile-wise (LE bn) With and Without AHD

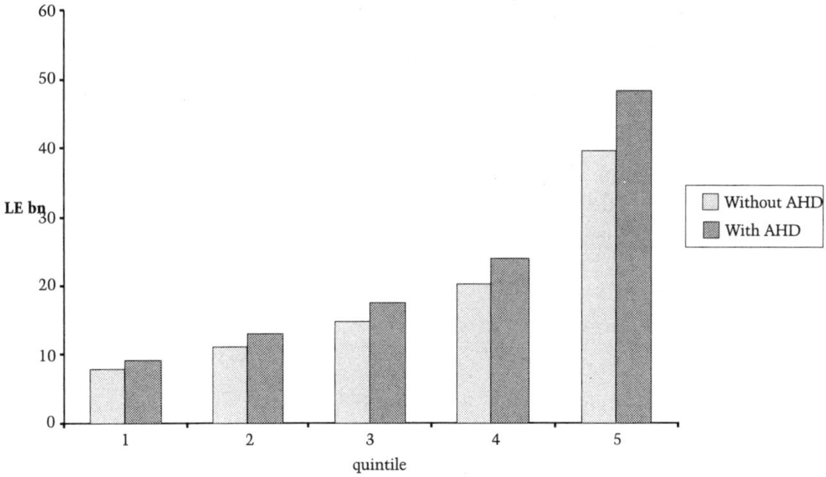

Conclusions

A few empirical conclusions can be drawn from this analysis.

- The high Aswan dam was, and is, a good investment, yielding significant annual net returns. The model analysis of net benefits is conservative, ignoring some benefits, which are significant but difficult to model (e.g., the elimination of major damages from periodic serious flooding).

- For agriculture, the existence of the dam allowed Egypt to pursue policies that distorted agricultural production, yielding a cropping pattern that favoured low-value crops that made inefficient use of both water and land. Reducing summer water and using the remaining water efficiently would yield increases in the value of agricultural production. With or without the dam, eliminating these distortions would greatly increase efficiency in Egyptian agriculture.
- Given the distortions in agricultural production, the largest gains from the dam arise from non-agricultural sources: hydropower, transportation, and tourism.

In terms of methodology:

- The use of unconstrained demand multiplier analysis to compute the benefits of a project such as a dam will tend to yield significant overestimates of benefits.
- Assuming that only labour is unconstrained, which should be a realistic assumption in many developing countries, yields a significant increase in estimated benefits compared to a model in which all factors are in fixed supply, but the estimates seem reasonable and are much less than those coming from a simple unconstrained demand-multiplier model.
- In estimating the impact of a large project such as the high Aswan dam, which affects a significant share of aggregate economic activity and has significant effects on many prices, a CGE model provides a good framework for analysis. The inclusion of water in the model, however, requires major extensions to the 'standard' CGE model.

References

Ali, Hesham M. and Mohamed R. Mahmoud (n.d.). *Determining Optimal Allocation and Cropping Pattern in Egypt by the Use of linear programming.* Ministry of Public Work and Water Resources, Government of Egypt.

Berck, Peter, George Goldman, and Sherman Robinson (1991). "The Use of Computable General Equilibrium Models to Assess Water Policies", in A. Dinar and D. Zilberman (eds.), *The Economics and Management of Water and Drainage in Agriculture.* Amsterdam: Kluwer Publishing Co.

CAPMAS (Central Agency for Public Mobilization and Statistics) (1996a). *Statistical Yearbook*. June. Cairo.

______. (1996b). *Harakat al-Intaj wat-Tijara al-Kharijiyya wal-Istihlak li-Ba'd as-Sila' hatta 'am 1994/95*. June.

______. (1998). *Statistical Yearbook*. June. Cairo.

Central Bank of Egypt (1995). *Annual Report 1994- 95.* Cairo.

______. (1998). *Annual Report 1997-98.* Cairo.

Hurst, H.E. (1952). *The Nile.* London: Constable

IFPRI (1998). *Egypt Integrated Household Survey (EIHS) 1997: Overview and Description of Version 2 Data Files*.

IMF. (1998). *International Financial Statistics Yearbook*. Washington D.C.

Kherallah, M., H., Lofgren, O. Badiane, P. Gruhn, and M. Mendoza (1998). *Wheat Policy Reform in Egypt: Adjustment of Local Markets and Options for Future Reforms.* Washington D.C.: International Food Policy Research Institute.

Lofgren, Hans, and Sherman Robinson (1999). "The Mixed-Complementarity Approach to Specifying Agricultural Supply in Computable General Equilibrium Models", in George H. Peters and Joachim von Braun (eds.), *Food Security, Diversification, and Resource Management: Refocusing the Role of Agriculture?*. pp.367-379. Brookfield, VT: Ashgate Publishing.

Lofgren, Hans, and Moataz El-Said (2001). "Food Subsidies in Egypt: Reform Options, Distribution and Welfare," *Food Policy* 26: 65-83.

Lofgren, Hans, Rebecca L. Harris, and Sherman Robinson (with assistance of Marcelle Thomas and Moataz El-Said) (2001). "A Standard Computable General Equilibrium (CGE) Model in GAMS", *TMD Discussion Paper* No. 75. Washington, D.C.: International Food Policy Research Institute.

Robinson, Sherman, Andrea Cattaneo and Moataz El-Said (2001). "Updating and Estimating a Social Account Matrix Using Cross Entropy Methods", *Economic Systems Research* 13: 47-64

Robinson, Sherman and Clemen Gehlhar (1996). "Impacts of Macroeconomic and Trade Policies on a Market-Oriented Agriculture", in L.B. Fletcher (ed.), *Egypt's Agriculture in a Reform Era*. Iowa: Iowa State University Press.

Shiklamanov, I.A. (2000). "Appraisal and Assessment of World Water Resources", *Water International* 25(1).

Thissen, M., and H. Lofgren (1998). "A New Approach to SAM Updating with an Application to Egypt", *Environment and Planning Series A* 30(11): 1991-2003.

Yohannes, Y., and H. Bouis (1999). "Unpublished Consumption and Elasticity Data for Egypt". Photocopy.

Annex A-8.1

Historical Nile Flows and Losses to Summer Agriculture, Navigation and Tourism

Table A.8.1. lists historical data on the Nile flood levels, summer flows, and associated losses to summer agriculture production in the following year, the reductions to navigations from low summer flows, and to tourism from low summer flow. Noting that no dam implied no hydropower. Assuming a 25 per cent decline in the electricity sector productivity captured the effect of no dam on hydropower across all historical runs.

Table A-8.1

Historical Nile Flows and Losses to Summer Agriculture, Navigation and Tourism.

Model Runs	*Flood Level*	*Summer flows*	*Summer agriculture*	*Navigation*	*Tourism*
Base	1.000	1.000	0.000	0.000	0.000
1	4.080	1.139	0.000	0.000	0.036
2	4.701	1.131	0.000	0.000	0.120
3	2.967	0.975	0.303	0.000	0.000
4	3.270	0.909	0.000	0.000	0.000
5	5.101	0.866	0.400	0.000	0.120
6	3.162	0.865	0.000	0.000	0.000
7	2.148	0.863	0.000	0.000	0.000
8	2.420	0.855	0.000	0.000	0.000
9	2.486	0.850	0.000	0.000	0.000
10	3.086	0.834	0.468	0.000	0.000
11	2.764	0.827	0.000	0.000	0.000
12	2.184	0.816	0.000	0.000	0.000
13	2.781	0.800	0.000	0.000	0.000
14	3.691	0.799	0.000	0.000	0.000
15	3.702	0.797	0.004	0.000	0.000
16	3.948	0.792	0.000	0.000	0.017
17	3.437	0.787	0.000	0.000	0.000
18	3.888	0.783	0.000	0.000	0.009
19	4.096	0.751	0.000	0.000	0.039
20	3.106	0.750	0.327	0.000	0.000
21	3.045	0.748	0.000	0.000	0.000
22	3.232	0.737	0.000	0.000	0.000
23	2.395	0.736	0.000	0.000	0.000
24	2.599	0.725	0.000	0.000	0.000
25	3.035	0.714	0.000	0.002	0.006
26	2.981	0.708	0.079	0.003	0.008
27	4.151	0.705	0.000	0.003	0.059

contd...

contd...

Model Runs	*Flood Level*	*Summer flows*	*Summer agriculture*	*Navigation*	*Tourism*
28	4.206	0.697	0.000	0.005	0.074
29	2.860	0.693	0.139	0.005	0.016
30	2.369	0.693	0.000	0.006	0.016
31	4.188	0.691	0.000	0.006	0.073
32	3.774	0.689	0.000	0.006	0.018
33	3.164	0.689	0.000	0.006	0.018
34	2.180	0.683	0.000	0.007	0.021
35	3.472	0.667	0.000	0.010	0.029
36	2.377	0.657	0.000	0.012	0.034
37	3.145	0.652	0.408	0.013	0.036
38	2.945	0.643	0.000	0.014	0.040
39	3.686	0.639	0.000	0.015	0.043
40	1.795	0.633	0.000	0.016	0.046
41	2.671	0.632	0.000	0.016	0.046
42	3.668	0.629	0.000	0.017	0.048
43	2.234	0.629	0.000	0.017	0.048
44	2.908	0.627	0.000	0.017	0.049
45	1.658	0.627	0.000	0.017	0.049
46	2.547	0.614	0.000	0.019	0.055
47	2.756	0.613	0.000	0.019	0.055
48	3.511	0.613	0.000	0.019	0.056
49	3.156	0.603	0.000	0.021	0.061
50	2.400	0.603	0.000	0.021	0.061
51	3.625	0.599	0.000	0.022	0.062
52	2.667	0.596	0.000	0.022	0.064
53	1.433	0.595	0.000	0.023	0.065
54	2.796	0.590	0.000	0.023	0.067
55	2.404	0.584	0.000	0.024	0.070
56	2.735	0.576	0.000	0.026	0.074
57	3.355	0.576	0.497	0.026	0.074
58	3.680	0.575	0.000	0.026	0.074
59	3.174	0.573	0.000	0.026	0.075
60	2.672	0.572	0.400	0.026	0.076
61	3.441	0.572	0.000	0.027	0.076
62	3.943	0.565	0.000	0.028	0.095
63	3.383	0.559	0.000	0.029	0.082
64	3.530	0.552	0.000	0.030	0.086
65	3.256	0.547	0.000	0.031	0.089
66	2.694	0.539	0.000	0.032	0.092
67	1.545	0.526	0.000	0.035	0.099
68	3.376	0.504	0.000	0.038	0.110
69	2.980	0.499	0.000	0.039	0.112
70	3.208	0.460	0.000	0.042	0.120
71	2.787	0.442	0.000	0.042	0.120
72	1.572	0.379	0.133	0.042	0.120

9 Sobradinho Dam and the Cascade of Reservoirs on the Sub-médio São Francisco River, Brazil

Monica Scatasta

Introduction and Overview

The premise of this study is that "a simple accounting for the direct benefits provided by large dams[...] often fails to capture the full set of social benefits associated with these services. It also misses a set of ancillary benefits and indirect economic (or multiplier) benefits of dam projects" (WCD, 2000). Nowhere does this seem more cogent than for the large dams and reservoirs located in the driest region of the São Francisco river basin, in Brazil's Northeast region. The São Francisco is the largest and one of the few permanent rivers in a region characterised by irregular rainfall patterns, high average temperatures and evaporation rates, and cyclical droughts that are occurring with increasing frequency—every five years or less—and persisting over more than one year. Despite the irregularity and concentration of rainfall and the fragility of the Northeast's soils, its vast interior was already occupied by the end of the XVIII century, fuelled primarily by extensive livestock production accompanied by subsistence farming. The persistence of traditional land use practices further degraded soils, limiting their capacity to retain rainfall. In spite of the rapid urbanisation process in coastal areas, the majority of people in the semi-arid interior still reside in rural areas, generally in dispersed centres with few economic alternatives.

The construction of the Sobradinho dam in 1973-1979 was expected to transform the region's economy, society, and landscape. With a surface of over 4,200 Km^2 and a water storage capacity of 34 billion m^3, the reservoir contributed not only to stabilising downstream flows for the reliable operation of a hydropower complex with a current installed capacity of about 10 GW, but also to provide water for large irrigation projects that are revolutionising agricultural production in the region (Nishizawa and Uitto, 1995). And yet, despite the existence of a number of success stories in the region, its socio-economic vulnerability appears to persist. This study analyses the direct and indirect economic impacts of the set of large dams in the medium and lower reaches of the São Francisco river. Its goal is to understand whether these impacts have, as some maintain, fallen short of expectations, and what factors have affected their magnitude, including those falling outside the scope of dam design and water use decisions. The study also highlights some of the emerging success stories, trying to reveal both their multiplier impacts and the lessons to be learned from them. Finally, it presents a brief overview of the social and environmental impacts of the selected dams.

The São Francisco River Basin: Hydro-climactic Characteristics and Water Use

The São Francisco river is 2,700 km long and its basin has a surface of 639.219 km^2 (7.5 per cent of the country's area) The basin comprises 504 municipalities in Minas Gerais, Bahia, Goiás, Pernambuco, Alagoas, Sergipe, and the Federal District, with a population of over 16 million. Brazil's Northeast macro-region accounts for 62.5 per cent of its area, over half of its municipalities and 36.3 per cent of its population. A large share of the Northeastern population in the basin resides in areas that fall either in the Drought Polygon or Semi-Arid region[1]—100 per cent in Pernambuco, over 80 per cent in Bahia, and about 70 per cent in Alagoas (Fibge, 1999 and Sudene, (1999) (Figure 9.1).[2]

With an annual available discharge of 64.4 BCM/year, the São Francisco represent 69 per cent of the available surface water in the

1. The Drought Polygon is a political and administrative region, legally recognised as subject to periods of prolonged drought, but with some areas with positive water balance. The semi-arid is one of Brazil's six climactic zones, defined based on its precipitation regime: average annual rainfall below 800 mm, irregular and concentrated in three months of intense and short rainfall episodes.

2. All Maps are modified versions of those in CODEVASF (2000). *Cadastro do Vale do São Francisco—2000*, available at *www.codevasf.gov.br* This section draws heavily on information contained in that document.

Figure 9.1 A

The Basin and the Country

Figure 9.1 B

The Basin, the Northeast Region and the Drought Polygon

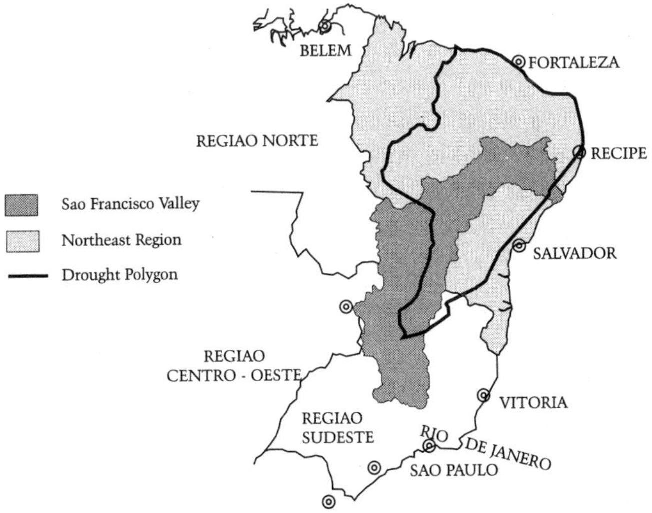

Northeast region[3] and 73 per cent of its guaranteed surface water availability. The large reservoirs in the Sub-Médio São Francisco, while contributing to evaporation losses, account for about 60 per cent (50.9 BCM[4]) of the region's surface water accumulation capacity. Despite its high overall average, however, the São Francisco river flow is plagued with significant fluctuations, with annual averages ranging from 1,768 to 5,244 m^3/s, and monthly values from 644 to 13,743 m^3/s.[5] This is due in part to the intermittent nature of many of its 168 tributaries. Most of the perennial tributaries are located in its upper reaches and the last significant one—the Rio Grande, in Bahia—merges with the São Francisco 1,178 km from its estuary. Most tributaries in the Drought Polygon are torrents whose regimes depend solely on rainfall patterns. Thus, the region between the MG-BA border and the city of Juazeiro represents 45 per cent of the basin area but contributes only 20 per cent of its annual flows.

The climate of the São Francisco Valley is characterised by high average temperatures, with very high evapotranspiration (896 mm/year on average), particularly in the Northern part of the Valley. The rainy season runs November through January, when 55 to 60 per cent of annual precipitation takes place, while June-August are the driest months. Rainfall decreases moving towards the riverbed and downstream. The highest annual rainfall levels (1,600 mm) take place at the source, while the lowest (350 mm) characterise the region along and between the studied reservoirs, then increasing again towards the estuary (Figure 9.2). Based on these characteristics, the basin can be divided into four regions (Figure 9.3). The studied dams are located in the Sub-Médio region, in Bahia and Pernambuco, characterised by semi-arid/arid climate (average annual temperature 27°C, evaporation 3,500 mm, and rainfall as low as 350 mm in the Sobradinho area), and a landscape dominated by scrublands. The Sub-Médio region accounts for 15 per cent of the basin's total population and 23.5 per cent of its rural population, with an urbanisation rate of 56.[6]

---

3. The region loses 88 per cent of its annual rainfall to evaporation or evapotranspiration. Only 8.6 per cent flows as surface water and 3.4 per cent infiltrates in ground-water aquifers
4. Sobradinho (34.1 BCM), Itaparica (11.8 BCM), Xingó (3.8 BCM) and Moxotó (1.2 BCM)
5. Data from *www.cbhsaofrancisco.org.br*
6. Elaboration of data from *www.cbhsaofrancisco.org.br*, the official site of the São Francisco river basin committee.

Figure 9.2

Rainfall and Climactic Zones

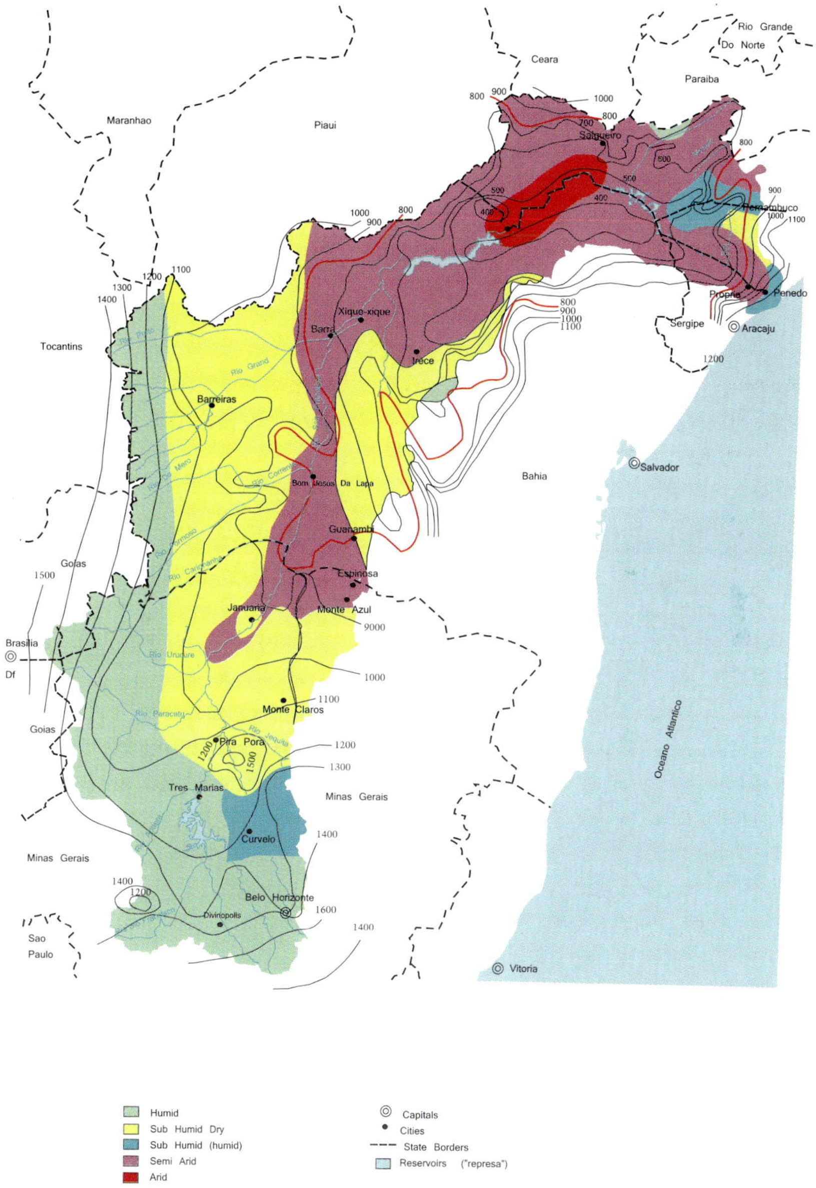

Figure 9.3

Regions and Main Tributabies

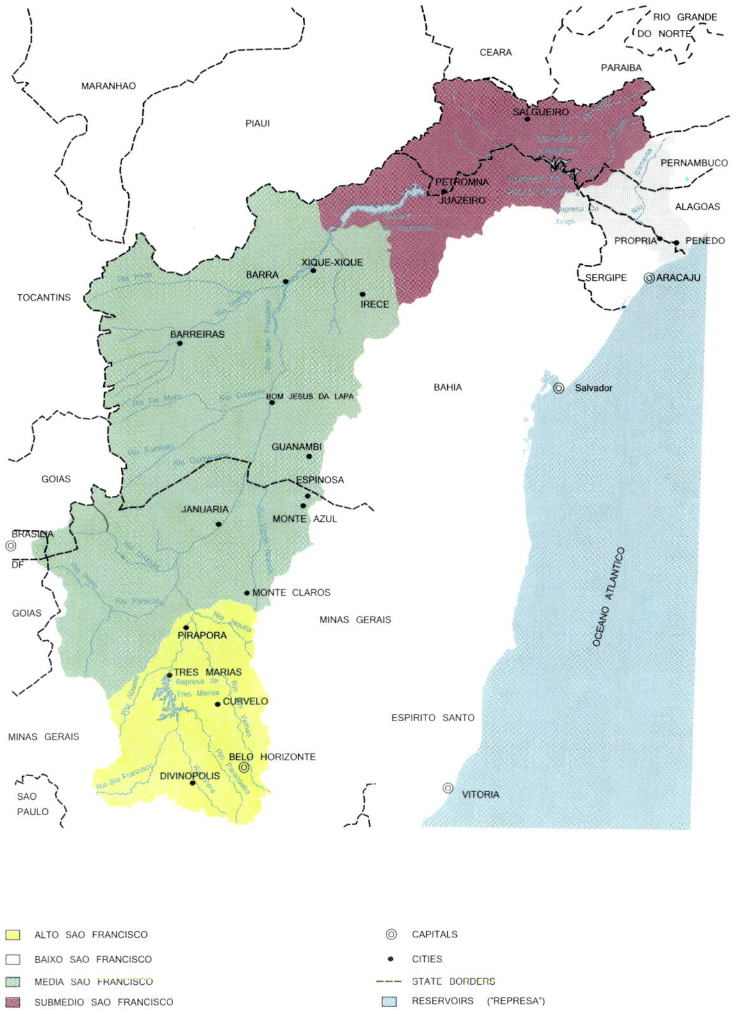

Critical hydrological events in the basin include flooding along the tributaries of the Alto São Francisco, and droughts primarily in the Médio and Sub-médio regions, where water is largely the limiting factor for human and animal life, and for productive activities. In recent years, droughts were responsible for declines in agriculture output of about 9 per cent, and for a 4.5 per cent reduction of GDP, affecting 12 million people and forcing 2 million workers to apply for emergency work programmes. Maia Gomes and Vergolino (1995) estimated supply and demand-induced impacts of droughts on agriculture, showing that the event often explained most of the difference between actual and trend GDP growth rates, and that this was not confined to the severe droughts of 1993 and 1983. Water availability and use in the regions are summarised in Table 9.1.

Table 9.1 shows that some areas are already facing critical demand to availability ratios. Irrigation demand is the largest water user, particularly in the basin's mid and lower reaches, where soils are poorer and those apt for irrigation are far from alluvial plains, requiring pumping. Tensions between water users are emerging with respect to quantitative, qualitative and inter-temporal uses. Increasing water scarcity, for instance, is reducing water bodies' capacity to dilute effluents from municipal and industrial users.[7] The main water use conflicts in the basin, however, appear to be between hydropower generation and irrigation, the primary user of water stored in the dams along the Sub-Médio São Francisco.[8] Their historic conflict is intensifying due to a plan to transfer water into and out of the basin—the "Transposição" of the São Francisco—to provide water for municipal supply, hydropower and irrigation, primarily to the Drought Polygon.

---

7. While water supply coverage is relatively good (80-94.4 %), but generally unreliable and of poor quality, particularly in the Drought Polygon and rural areas, wastewater collection coverage is much lower and more unevenly distributed—10 to 45 per cent in the Sub-Médio, and less than 20 per cent in the Baixo region. Wastewater treatment coverage is lower, but close to the national average (20%). Industrial pollution is a lesser problem thanks to good monitoring and enforcement by state-level environmental protection agencies. (Data from *www.cbhsaofrancisco.org.br*).
8. Relevant trade-offs also exist between hydropower and flow regulation, navigation and fish movement. This matters given the impressive potential for commercial fishing in the basin. The reservoirs also provide opportunities for recreational uses, eco-tourism and recreational fishing, but their development, is just incipient.

Table 9.1

Water Availability and Demand in the São Francisco Basin and Its Regions

Sub-basin	*Area (km²)*	*P (mm)*	*E (mm)*	*Q (m3/s)*	*Demand (m³/s) % Total*					*Irrig. Demand /Flow (%)*	
					Human	*Irrig.*	*Animal*	*Industr.*	*Total*		
Alto	111,804	1,402	1,051	1,013	16	22	1	18	57	40.35	6
Médio	339,763	1,111	952	1,773	6	64	4	3	77	87.18	3
Sub-médio	155,637	695	619	375	3	63	1	6	73	87.67	20
Baixo	32,013	842	694	217	3	11	1	2	17	75.00	7
Total Basin	639,219	1,036	896	2,850 1/	28	160	7	29	224	74.55	8

Notes: P – precipitation, E—evapotranspiration, Q—average annual flow

1/ The overall average is smaller than the sum of the sub-basins' due to evaporation losses from flow regularisation reservoirs

Source: *www.cbhsaofrancisco.org.br*

Direct Impacts of the Cascade of Reservoirs

Hydropower Development along the Sub-Médio São Francisco[9]

Hydropower was the primary goal behind the construction of the large reservoirs along the Sub-Médio São Francisco, and plays a crucial role in the management of the basin's waters. The basin's hydropower potential is estimated at 26,300 MW, of which 10,533 (15 per cent of Brazil's installed capacity) have already been developed, with 33 power stations currently in operation, 9 of which are along the São Francisco river (Figure 9.4). These provide the backbone of power supply to the North/ Northeast electric *polo*, which was integrated with the 1981 inauguration of the Boa Esperança/Imperatriz transmission line linking CHESF's with Eletronorte's lines.[10]

9. Data from *www.energiabrasil.gov.br*, *www.chesf.gov.br*, *www.portalsf.com.br*, where not otherwise indicated.

10. The role of the transmission in the Brazilian hydroelectric system is of major importance, as it permits the optimisation of generation, increasing overall energy output by transferring surplus energy to regions with unfavourable hydrological conditions. The optimised operation of the Brazilian system results in an increase of about 30 per cent in energy generation, compared to the production that would be obtained through the operation of the plants without the coordination of a centralised dispatch. In the Brazilian electric power system, the market supply criterion is based upon meeting average energy requirements, attributing a lower priority to peak supply. The system adopts a criterion based on the supply of guaranteed energy with a certain deficit risk. Currently the maximum annual deficit risk considered for expanding the electric power generation is 5 per cent. (Ventura Filho, 2002)

Figure 9.4

Main Power Infrastructure in the Basin

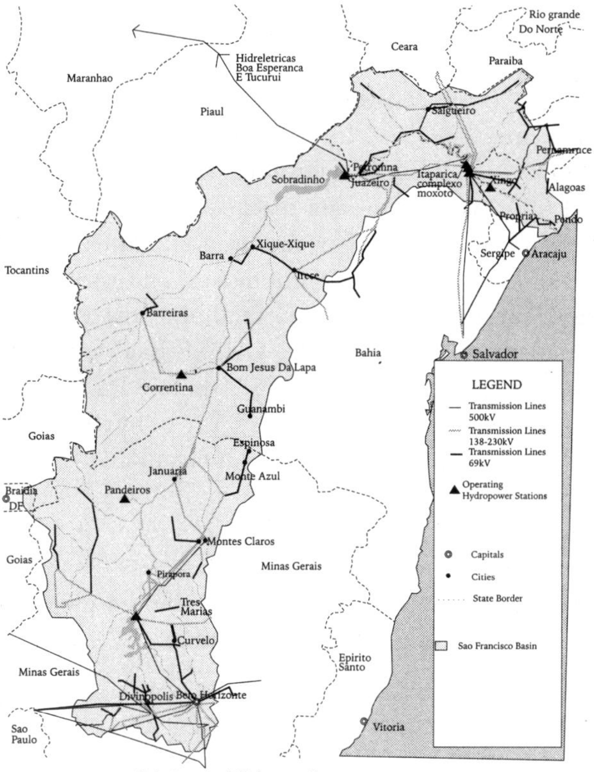

Main Operational, Hydropower Plants

	Generation Potential (MW)	Water Storage (Billon m3)
Tres Mari as	396.0	19.3
Sobradi Nho	1,050.0	34.1
Itapari Ca (**)	1,500.0	11.8
Complexo Moxoto(*)	4,283.0	1.2
Xingo	3,000.0	3.8

(**) Now Luis Gonzaga
(*) INCL. Paulo Afonso

After one hundred years of studies and pilot projects,[11] and the creation of CHESF (the São Francisco Hydroelectric Company) in 1948, Paulo Afonso I was inaugurated in 1955 and in 1959 the first transmission lines started providing power to Northeastern state capitals. In 1961 and 1971, the Paulo Afonso (PA) II and III stations were activated, taking capacity to 1,500 MW. Increasing power consumption required CHESF to build regularisation reservoirs that

11. In 1859, the emperor Dom Pedro II ordered the realisation of studies on its hydroelectric potential, and the first pilot station was built in 1913.

would ensure flow reliability. The Moxotó[12] reservoir (100 km^2, 1.2 BCM) was built 4 km upstream of PA to provide weekly inflow regularisation and additional capacity (440 MW). Pluriannual flow regularisation was also needed to increase minimum guaranteed dry season flows and ensure the viability of a fourth station in the PA complex, leading to the construction of Sobradinho dam, about 470 km upstream of it, in 1973-1979. Its objective was to maintain a minimum flow of 2,060m^3/s, as well as additional generation, with an installed capacity of 1,000 MW. [13] Its costs reached US$ 987 million, including the cost of the navigation gates of the head structures for irrigation, but excluding a full assessment of environmental and social impacts. A few months before the inauguration of Sobradinho, the PA IV plant (2,400 MW) was inaugurated. Meanwhile, CHESF had started building the Itaparica reservoir and power station (834 km^2, 11.8 BCM, 1,500 MW). Its inauguration was planned for early 1981, but financial problems produced delays that, together with an unfavourable hydrological cycle, caused energy rationing in the Northeast in 1987-88. The latest addition to the system was the Xingó station, downstream of PA, in the Baixo São Francisco, which started operating in 1994 and reached total capacity (3,000 MW) in 1997. The US$ 3.1 billion project accounts for 25 per cent of installed capacity in the Northeast.

In the late 1970's, Brazil could rely on various alternative primary energy sources, including coal, uranium. On the other hand, oil and natural gas reserves were too low, and their import price too high, for a large-scale conventional thermoelectric generation expansion programme. In 1994, ELETROBRÁS estimated the relative competitiveness of various sources (Table 9.2).[14] Today, only 33 per cent of Brazil's hydroelectric potential has lower costs than other sources, although two thirds remain competitive. At the time of their construction, however, the studied dams were provided electricity at costs (around 1,000 US$/kW or less—roughly 20 US$/MWh) (Ventura,

12. Renamed Apolônio Sales, in homage to CHESF's founder.
13. More recent sources set its installed capacity at 1,140 MW at *ww.portaldosaofrancisco.hpg.ig.com.br*
14. The study states that all costs (investment, operation/fuel and maintenance, and environmental) were considered. Benefits were calculated for each plant operating under optimised conditions within the interconnected system. A 10 per cent annual discount rate and a useful life of 50 years for hydro and 30 years for thermal plants was assumed.

Table 9.2

Brazil: Competitiveness of Primary Sources for Electric Power Generation

Source	*Cost (US$/MWh)*
Natural Gas 1/	40 to 50
Biomass/Sugarcane Bagasse	40 to 80
Oil By-Products	50 to 60
Mineral Coal	50 to 65
Nuclear Power 2/	60 to 70
Hydroelectric Power	33% : less than 403 39% : between 40 and 70 28% : over 70

Note: 1/ Recent bids for imported natural gas fired thermal generation point to costs in the range of 30 US$/MWh

2/ ELETRONUCLEAR, the Brazilian Nuclear Engineering Company and Utility, foresees some reduction in these.

Source: ELETROBRÁS (1994)

2002) that are at least 50 per cent lower than those of the currently cheaper alternative, and were likely more competitive at a time of higher fossil fuel prices and more expensive imported technology.

Today, the hydropower plants along the Sub-Médio São Francisco represent over 97 per cent of CHESF's 10.7 GW installed capacity, the largest of any power company in Brazil, representing over 12 per cent of the country's capacity and accounting for 80 per cent of power generation in the Northeast. Through its power generation, transmission and distribution activities, CHESF serves primarily eight Northeast states benefiting over 40 million people in an area of influence of 1.2 million km^2 (about 14.3 per cent of Brazil's area). Beneficiaries, however, are mainly located in urban areas, as severe imbalances still characterise access to electricity in the region, where most villagers and small holders lack many basic requirements, and less than half of rural households have access to electricity (World Energy Council, 2003). In addition, the large dams and the interconnected nature of Brazil's system did little to prevent the 2001 energy crisis, due to a combination of natural causes and under-investment in generation, as several years of below average rainfall had left reservoirs over 70 per cent depleted, while installed capacity expansion had not kept up with the steady growth of electricity demand since 1980.[15] The crisis was overcome primarily thanks to consumers' response to the rationing programme that mandated 20 per cent consumption reductions across the board.[16]

Irrigated Agriculture in the São Francisco River Basin

The regularisation of the São Francisco river benefited the agriculture sector, providing reliable sources of irrigation water that supported irrigation expansion. In the late 1960's, the Federal government launched the Multi-year Irrigation Programme to spur socio-economic development in the semi-arid region by increasing in its agriculture output and productivity.[17] Since then, the Northeast has been the theatre of massive public sector irrigation projects.[18] Two public implementing agencies operate in the region: CODEVASF (the São Francisco River Valley Development Company), created in 1974 to lead irrigation development in the basin,[19] and DNOCS (the National Department of Anti-Drought Works), in charge of drought relief and irrigation in the rest of the region. Table 9.3 shows the evolution of

---

15. Pêgo Filho *et al.* (2001), Watts and Aritzia (2002), and information at *www.iie.org*, *www.fe.doe.gov*
16. Residential consumers exceeded such targets by eliminating wastes and replacing inefficient household appliances, while industry relied on existing own generation capacity, often in the form of cogeneration. Reforms in the 1990s introduced competition in generation. While not a cause of the crisis, their lack of clarity and investors' risk aversion delayed the construction of gas-fired plants after the opening of the Bolivia-Brazil pipeline. The 2002 Northeastern Water Infrastructure Programme called for CHESF's restructuring (avoiding its privatisation), the consolidation of the Interconnected System and the installation of thermoelectric plants close to large markets.
17. The programme focused on structural deficiencies of Northeastern agriculture such as vulnerability to climate and low productivity of food crops; conflicts over factors; supply disequilibria that caused deficits of grains, basic foodstuff and export goods; subsidies; and repressed demand for a more diversified production.
18. By 1970, the Northeast's share of Brazil's irrigated area had passed from 0.2 per cent to 15 per cent, and has only inched upwards since then. However, it was only after 1970 that the public share of irrigated area in the region surged, going from 7 per cent to 29 per cent by 1995 (SUDENE, 1997), and *www.codevasf.gov.br*). This seems to contradict the statement that public irrigation was needed because higher irrigation costs in the region would limit private investment, which carried out the bulk of irrigation development in the rest of Brazil
19. Substituting the Superintendência do Vale do São Francisco (SUVALE), which in 1967 had replaced the Comissão do Vale do São Francisco (CVSF), created in 1948.

total and public irrigated areas in the basin. CODEVASF estimated the region's untapped irrigation potential at 1.5 million ha.[20]

Table 9.4 shows recent estimates for current and expected irrigated areas in different parts of the basin. The areas directly affected by the

Table 9.3

Evolution of Total and Public Irrigated Area: São Francisco

Period Until	*São Francisco ('000 ha)*	*Yearly growth rate %*	*CODEVASF ('000 ha)*	*Yearly growth rate %*	*Public as % Total*
1950	0.1		0.1		100.0%
1960	10.8	59.7%	1.1	27.1%	10.2%
1970	60.2	18.7%	2.3	7.7%	3.8%
1975	88.0	7.9%	12.1	39.4%	13.8%
1980	144.5	10.4%	27.0	17.4%	18.7%
1985	205.9	7.3%	47.9	12.1%	23.3%
1990	232.6	2.5%	67.0	6.9%	28.8%
1995	300.0	20.0%	72.8	1.4%	24.3%
1999	333.0	2.6%	122.0	15.2%	36.6%

Sources: SUDENE (1999), CODEVASF (1989) and (1999), and author's calculations

studied dams are those located between Sobradinho and Xingó, which represent roughly 47 per cent of the basin's irrigated area.

Whether public or private, irrigation appears to have significant impacts on agricultural productivity and profitability, and on employment creation, in the Northeast. Considerably higher yields can be obtained in irrigated agriculture than in rainfed farming (Table 9.5).

Irrigators also tend to have higher and more stable net revenues: US$ 2,500-5,000/ha/year, and up to US$ 5,000 in public systems, *versus* US$ 260/ha/year for rainfed farmers. Irrigated agriculture in the region is also more labour-intensive than in the rest of Brazil, generating employment at lower costs than other sectors (Table 9.6) and providing higher salaries than rainfed agriculture.

20. This is less than 5 per cent of the soils apt for irrigation and less than 20 per cent of those for which surface water is available, due to considerations of soil distance and elevation from water sources, and of multiple water uses. Assuming consumption of 20,750 m^3/ha/year and a 30 per cent return flows, PLANVASF estimated that a smaller planned expansion of 600,000 ha would require 11.7 billion m^3/year (13% of the São Francisco's average annual flow). This would necessitate technological improvements and the regulation of the São Francisco and its tributaries

Table 9.4

Estimated Irrigated Area (in hectares) in the San Francisco River Basin

Year	*Upstream from Sobradinho*	*Between Sobradinho and Itaparica*	*Between Itaparica and Xingó*	*Downstream from Xingó*	*TOTAL São Francisco*	*LARGE DAMS AREA*
1999	162,407	149,619	6,885	14,399	333,310	156,504
2000	166,305	153,210	7,050	14,745	341,310	160,260
2001	170,203	156,801	7,216	15,090	349,310	164,017

Source: ANA/GEF/UNEP/OAS (2002).

Table 9.5

Selected Crop Yields, Rainfed versus Irrigated Agriculture (t/ha)

Crop	*Region*	*Rainfed*	*Irrigated*
Corn	Northeast	3.1	4.0-4.2
	Brazil	2.8	4.8
Beans	Northeast	0.5	1.0-1.6
	Brazil	0.7	1.5
Rice	Northeast	1.1-1.7	2.6-3.4
	Brazil	2.5	4.8

Source: MMA/SRH (1999).

Perhaps more importantly, irrigation and the programmes that have recently accompanied its expansion have encouraged the shift to higher value crops in a number of areas of the basin, and these crops are gaining growing access to export markets. Fruit production is developing at a vertiginous pace, covering nearly 100,000 ha, with recent average increments of 9,000 ha/year. Output growth has been particularly significant for bananas, coconuts, guava, mangoes, and grapes. Fruit producers proved particularly apt at gaining access to new markets for their products and have shown a great capacity to react to market signals.

This shift was facilitated by the link between agricultural development and agribusiness that was explicitly pursued by SUVALE and CODEVASF. Public irrigation was developed as part of the creation of regional development districts (*pólos*) (Figure 9.5), based on the identification of municipalities and agricultural products presenting the potential for the expansion of integrated agro-industrial complexes. Their objectives were to (i) stimulate private participation in irrigation

Figure 9.5

Public Irrigation and Regional Development Districts

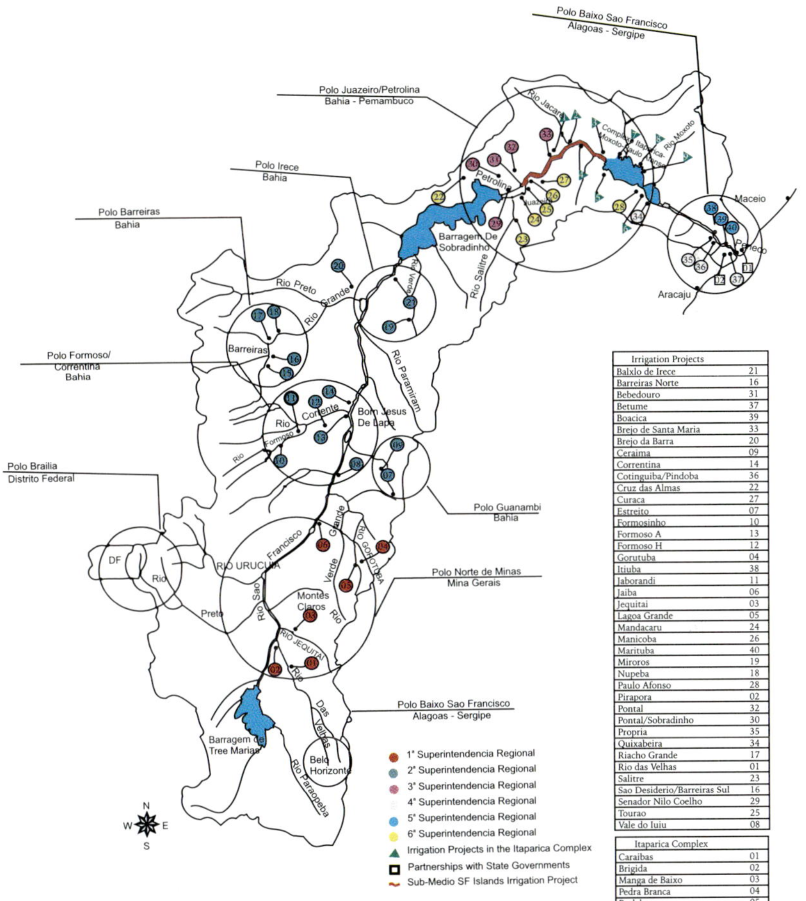

Irrigation Projects	
Balxlo de Irece	21
Barreiras Norte	16
Bebedouro	31
Betume	37
Boacica	39
Brejo de Santa Maria	33
Brejo da Barra	20
Ceraima	09
Correntina	14
Cotinguiba/Pindoba	36
Cruz das Almas	22
Curaca	27
Estreito	07
Formosinho	10
Formoso A	13
Formoso H	12
Gorutuba	04
Itiuba	38
Jaborandi	11
Jaiba	06
Jequitai	03
Lagoa Grande	05
Mandacaru	24
Manicoba	26
Marituba	40
Miroros	19
Nupeba	18
Paulo Afonso	28
Pirapora	02
Pontal	32
Pontal/Sobradinho	30
Propria	35
Quixabeira	34
Riacho Grande	17
Rio das Velhas	01
Salitre	23
Sao Desiderio/Barreiras Sul	16
Senador Nilo Coelho	29
Tourao	25
Vale do Iuiu	08

Itaparica Complex	
Caraibas	01
Brigida	02
Manga de Baixo	03
Pedra Branca	04
Rodelas	05
Gloria	06
Ico-Mandantes	07
Apolonio Sales	08
Barreiras	09

Partnership Projects	
Pindorama	01
Plato de Neopolis	02

Table 9.6

São Francisco River Basin: Cost of Creating a New Job, by Activity

Activities	*Cost (US$)*
Irrigation	2,000–20,000
Wood Processing/Furniture	9,900–128,400
Hotel Services	12,300–140,800
Constructions	15,200–227,400
Steel	15,600–725,100
Fertilizers	29,800–193,100
Machinery	38,900–169,600
Pulp and Paper	46,200–669,100
Auto Industry	47,300–127,000

Source: CODEVASF (1996a)

expansion, as their scale would reduce input costs, attract specialised workers, improve trading opportunities and favour the introduction of technological innovation; (ii) create employment and reorient migration; (iii) turn the Northeast into an exporter of fruits, vegetables and seeds; and (iv) develop small and medium-sized urban centres to reduce growth and congestion of large coastal cities. Support for district development also entailed urban development interventions, investment in energy and transportation, and the formation of specialised labour.

Over time, the *pólos* turned into a motor of growth for the studied region.[21] The most successful is the one around the cities of Petrolina (PE) and Juazeiro (BA), located along and downstream of the Sobradinho reservoir. Its development started in the 1960's, with the first irrigators, mainly small farmers, settling in 1968. CODEVASF later established projects that also offered medium to large plots to corporate farms. Agricultural production rose substantially, particularly since the late 1980's, when tomatoes, melons, cotton, grapes, and mangoes became major crops.[22] Agro-processing played a role in its rapid growth, with a special mention owed to fruit pulp and mango juice; wine and vinegar;

---

21. For recent discussions of their structure and performance, see Accarini (2002) and França and Maia (2002).

22. Nishizawa and Uitto (1995) report that, in Juazeiro, the area planted with tomato went up from less than 900 ha in 1984 to 4,700 ha in 1988, and that planted to mangoes increased from 20 ha in 1986 to 240 ha in 1987.

tomato paste and canned vegetables; tobacco manufacturing; seeds and seedlings; cotton processing; sugar and alcohol; honey and sweets. The district is now the major producer of table grapes in the country, accounting for 90 per cent of grape and 80 per cent of mango exports, primarily to Europe and the USA, for a total of US$ 36.5 million in 1999.[23] Overall, 40 per cent of the pólo's fruit crop output is exported, for a value that reached US$ 65 million in 1997, representing 40 per cent of Brazil's fruit exports (Lima and Miranda, 2001).

Recent successes with development districts, however, followed a lacklustre start, and advancements were achieved only after a marked change in course.[24] The integrated rural development projects undertaken in the Northeast between 1975 and the mid-1980s,[25] were complex interventions that targeted low-income producers with production services and subsidies, and provided selected focus areas with physical and social infrastructures. Public irrigation interventions consisted mainly of medium- and large-scale, area-specific (*enclave*) projects, with substantial investments in water storage or river diversions. The projects lacked clear priorities, pursuing a mix of production, rural poverty alleviation and drought control objectives, that simply assumed the existence of a link between increased agricultural output by small farmers and poverty reduction. Economic efficiency was often sacrificed, and the projects did not focus on training, marketing activities or the switch to high value crops, nor did they rely on analyses of cross-sector and environmental impacts of irrigation. In the end, the projects were marred by complexity and inflexibility, large implicit subsidies, an incapacity to increase small farmers productivity, top-down design and implementation, lack of counterpart funding, and a failure to identify the key constraint to rural development at the time: an economic policy environment that penalised agriculture to promote import-substituting industrialisation.

The new generation of projects undertaken in the late 1980s and early 1990s[26] abandoned the pre-existing format of enclave-based

23. VALEXPORT 2000 data, reported in Correia *et al.* (2001).
24. Operations Evaluation Department (1991).
25. The World Bank financed 42 per cent of their total cost—for about US$ 1.4 billion.
26. These projects also received extensive World Bank support.

integrated rural development, and emphasised regional programmes with a sub-sector focus.[27] These new projects were accompanied by a comprehensive analysis of the agricultural sector and of the role of irrigation, embodied in the 1988 National Irrigation Review. Most fared better than the previous generation, placing more emphasis on linking project areas with markets and providing support to private irrigation development, while maintaining public irrigation mainly in the form of 'mixed' private-public schemes. Poverty reduction was pursued through separate State-level Rural Development Projects, which provided targeted investments deemed more effective at reaching the poor, such as investments in education and health, and employment generating activities. In the 1990's the National Irrigation Programme (PRONID) combined elements of the previous geographic and sector approaches, promoting integrated irrigation planning and management by coordinating public and private sector interventions within comprehensive development plans for each *pólo*. Public investments and government support focused mainly on the *pólos*, while irrigators outside the districts received credit support and technology transfers. Within each district, care was taken to assign responsibility for public infrastructure and complementary works to the appropriate level of government. As reforms in economic policies and water management progressed in the second half of the decade, so did the concept of integrated development underlying the *pólos*. This was revised, based on local development models promoted in the Federal programme *Brasil em Ação*, which called for better integration and partnerships between the public and private sector and civil society. The original *pólos* depicted in Figure 9.5 were thus consolidated based on an updated analysis of their development potential and integration with this programme.[28]

Irrigation management institutions need further integration with Brazil's water resources management framework, which has focused on integrated planning by participatory basin-level institutions since the promulgation of the National Water Law 9.433/97. Irrigation Law 6.662/79 is now largely outdated, due to the evolution of the country's

27. World Bank (1994)

28. Alagoas milk-production *pólo* (4146 km2, 265000 hab.), Petrolina/Juazeiro (24385 km2, 505600 hab.), Western Bahia (46255 km2, 225400 hab.) and Northern Minas (12607 km2, 176400 hab.). (Source: Banco do Nordeste, 1998).

economic and institutional framework, and hinders the harmonisation of actions in the irrigation sector with river basin planning and rural development initiatives. A Draft Law (P.L. 229) was presented to Congress in 1995 by the Special Senate Commission for the São Francisco Valley Development, but is still pending. Recently, CODEVASF's actions were integrated in a basin plan, based on an assessment of water resources availability and vulnerability, and of their multiple uses.[29]

The development of the *pólos* served by water regulated by the studied dams can be seen as one of their direct outputs. However, their focus on food processing as the strategic complement of agriculture may blur the distinction between direct and indirect benefits. The next section focuses on the factors that facilitated or hindered the realisation of indirect benefits.

Indirect Impacts: Driving Forces and Limiting Factors

This section has two objectives. One is to provide an overview of socioeconomic development that accompanied and followed the construction of the dams on the Sub-Médio São Francisco, highlighting some of the Northeast's persisting vulnerabilities, which many see as an indication that the dams failed to deliver what they promised. The second is to emphasise some of the contextual factors that might have affected their capacity to produce multiplier effects. The capacity of a dam to generate real indirect and induced benefits depends largely on how its outputs build on the strengths and reduce the weaknesses of the regional context. It is only by recognising driving and constraining factors, and by designing the project and ancillary programmes accordingly, that a dam's potential multiplier effects might be realised.

Between 1960 and 1994, GDP in the Northeast increased almost five-fold, while population and GDP per capita doubled. Table 9.7 shows, however, that early rapid increases in GDP were followed by stagnation between 1980 and the launch of the 1994 Plano Real reforms.

29. This plan includes the intra- and inter-basin transfers discussed earlier. While their discussion lies outside the scope of this study, but the database and model presented subsequently can help shed some light on their potential benefits.

The table highlights a tendency for GDP growth rates in the region and in Brazil to move in step. This has traditionally been interpreted as evidence of the region's integration with the country's economy. Recent studies, however, argue that this is true only for manufacturing, and

Table 9.7

Average GDP Growth Rate: Northeast Region and Brazil

Period	*Northeast*	*Brazil*	*Period*		*Northeast*	*Brazil*
1960/93	5.5	5.6	Adjustment Period	(1960-67)	4.4	3.9
1970/93	5.0	4.1	'Miracle' Boom	(1968-73)	7.2	10.9
1980/93	5.5	2.1	Oil Crisis	(1974-80)	7.4	6.8
1970-79	8.1	8.3	Debt Crisis	(1981-83)	2.6	-2.2
1980-89	4.4	2.7	False Recovery	(1984-86)	10.1	7.0
1990/93	-0.7	1.5	Hyperinflation	(1987-93)	-0.5	0.5

Source: Maia Gomes and Vergolino (1995).

that this 'integration' needs to be better characterised. The early surge in regional GDP was greatly influenced by fiscal incentives for the installation of industries in the region during the years of import-substituting industrialisation (ISI) policies, starting in the mid-1960's. Modern industries tended to locate in coastal areas and have few links with the interior and little focus on local markets. They appear to mirror the Southeast industrial complex, from which they depend heavily both as a source of capital and technology and as a market for their output, reducing intra-regional economic linkages.[30] Since the 1960's, the external and inter-regional trade balances of the region have shown deficits exceeding 10 per cent of GDP.[31] Regional absorption has traditionally exceeded output, with the difference financed largely by public transfers to states and municipalities, by fiscal incentives, or through the social security system. In rural areas and smaller cities, this represents the most important income source for lower income populations, representing up to 40 per cent of internal product in the Semi-Arid region. The region's continuing dependence on income

30. Guilhoto (1998), Guilhoto *et al.* (2000), Soares and Pereira (2000) reach these conclusions using inter-regional input-output models.

31. No data exists on service trade, but their inclusion is unlikely to tilt the balance, despite a recent boom in tourism.

transfers from the rest of the country is a reason for concern, as the availability of resources is dwindling and local governments are increasingly expected to achieve financial independence. The heavier impact that the 1987-1993 macroeconomic instability had on the Northeast may depend in part on the progressive waning of public interventions that played a crucial role in inducing and stabilising the region's economic growth in previous periods (Maia Gomes and Vergolino, 1995).

Another source of early GDP growth was the large investments in irrigated agriculture starting in the mid-1970. These, however, only partially compensated for the burden on the agriculture sector represented by the distortions introduced by ISI policies, which imposed an implicit tax on agriculture through the combination of overvalued exchange rates and protected prices for domestically produced agricultural inputs. Partially to compensate for this, subsidised rural credit was introduced, creating further distortions and almost exclusively benefiting medium-large producers, while millions of small farmers remained at the margin due to excessive collateral requests and bureaucratic rigidities. This contributed, at least initially, to worsening income concentration. In the absence of broader reforms aimed at removing the bias against agriculture of ISI policies, the large irrigation projects of the 1970's "made very little contribution towards solving the deep-seated agrarian problems of North-East Brazil," (Hall, 1978). The government started modifying its economic policies in the late 1980s, introducing trade liberalisation and deregulation in key sectors, including some agricultural sectors. It simultaneously revised agricultural price support policies and reduced the availability of official subsidised credit, encouraging reliance on internally generated funds and private lending at market interest rates.[32] The positive impacts on agriculture expected from a reduction in implicit taxation, however, were dampened by the subordination of these interventions to Brazil's macroeconomic objectives, due to the parallel difficulties faced by the stabilisation process. Regions and crops were affected differently by the interactions between macroeconomic and sectoral reforms, and positive

32. Which rose dramatically in real terms when stabilisation policies started having an impact on inflation.

impacts on the sector were slower and less widespread than expected (See Gasques, 2001, Gasques and Villa Verde, 1995). The net impact of three decades of, at times conflicting, public interventions in the Northeast was a dramatic change in the structure of the region's economy. From 1960 to 1994 the share of agriculture in GDP declined from 30.5 per cent to 15 per cent, while that of industry increased from 22 per cent to 27 per cent and services went from 47.5 to 58 per cent.

Economic stagnation and the decline of agriculture as a share of GDP, however, did not affect all areas of the Northeast equally. In the São Francisco basin the decline in agriculture share was less marked, due primarily to a greater than average expansion of capital intensive, irrigated agriculture. Irrigated areas, in addition, generally fared better than the rest of the Northeast region. Table 9.8 shows estimates of GDP and its evolution for four Northeast sub-regions—metropolitan areas, the semi-arid, irrigated areas, and the remaining *'zona da mata'*. While stagnation characterised the Semi-arid, which shrank from 43 per cent of the region's GDP in 1970 to only 15 per cent in 1992, metropolitan regions grew substantially, passing from 25 per cent to 50 per cent of regional GDP, driven by incentives to industry in the first

Table 9.8

Northeast and Sub-regions: GDP Estimates 1970-1992 (Million 1992 US$)

Sub-Region	*1970*	*1975*	*1980*	*1985*	*1988*	*1992*
Metro Regions	5,165	9,607	16,286	23,288	29,065	32,081
Mata	5,774	9,270	12,120	15,863	14,044	19,824
Semi-Arid	8,457	12,629	16,263	15,958	20,193	9,350
Irrigated	191	370	556	845	886	1,122
Total NE	19,587	31,876	45,225	55,954	64,188	62,379
(Index 1970 = 100)						
Metro. Regions	100.0	186.0	315.3	450.9	562.7	621.1
Mata	100.0	160.6	209.9	274.7	243.2	343.3
Semi-Arid	100.0	149.3	192.3	188.7	238.7	110.6
Irrigated	100.0	193.7	291.1	442.4	463.9	587.4
Total NE	100.0	162.7	230.9	285.7	327.7	318.5

Source: Maia Gomes and Vergolino (1995)

part of the period and by a boom in services in recent years. But it is the impressive record of irrigated areas that is of interest here, as they presented the same level of dynamism as metropolitan regions. While they doubled in their share of regional GDP, irrigated areas still represent a very small portion of the region's economy (2 per cent). This is nonetheless impressive given their very small area relative to the region as a whole.

The success of irrigated areas in the Northeast has been linked, in large part, with the switch to export-oriented, higher-value crops, and their link with agribusiness.[33] Thus, in the case of irrigation, the dams in the Sub-Médio São Francisco seem to have produced indirect and induced benefits. In addition, the negative distribution impacts of early irrigation and rural credit subsidies appear to have given way to a much different picture, at least in the development districts. Recent analyses show that irrigation development targeting medium and large producers generated significant positive spillovers for smaller farmers, as well as landless people employed in agriculture, and that positive indirect impacts exist for urban areas of municipalities with irrigation *versus* without irrigation in the Northeast (Rodrigues: 2001, Vergolino and Vergolino: 1997, Bonelli: 2001). Given the small area affected by irrigation linked with the large dams, however, their impacts may not have been substantial for the Northeast region as a whole. Their greatest value, however, may have resided in the success stories represented by the second and third-generation interventions in development districts, and in the replicability of the lessons learnt from them.

Too little scaling up seems to be taking place, however. Despite a recent upturn in the region's economic performance, its sustainability will depend on how effectively the Northeast addresses its environmental, socio-economic, scientific, technological, political and institutional sources of vulnerability. The sectoral and isolated interventions that characterised investment in the Northeast, while contributing to structural change, did little to ensure sustainable social development in the region. Despite recent advancements in a number of

33. For a discussion of current conditions of the agriculture/agribusiness productive chain see Haugenauer *et al.* (2001) and Prochnik and Haguenauer (2001).

indicators, income gaps kept widening.[34] More than 50 per cent of families in the region—over 22 million people, of which 12 million in rural areas—live below the critical poverty line. The poor of the rural Northeast represent 63 per cent and 32 per cent of the country's rural and total poor, accounting for 9 per cent of total population, but receiving less than 1 per cent of national income. These results extend beyond the arid interior to the rural Northeast as a whole. Urban areas in the region are also affected by high poverty incidence, which reaches 29 per cent in metropolitan areas and 41 per cent in non-metropolitan areas. Based on a poverty index linked with the capacity of families to get adequate nutrition, the Northeast municipalities of the São Francisco basin present an alarming picture. Table 9.9 compares them to the national average (NA) of 24.4 per cent per municipality. Virtually no municipality fared better than the national average, and almost all presented a poverty index of over 40 per cent.

As a result, the exodus from the rural Semi-Arid often represented an inter- and intra-regional poverty transfer. In the 1960s the net flow of migrants reached 4.1 million, 42 per cent of which moved to urban centres in the region, while the majority moved to other regions, primarily urban centres in the Southeast. In the 1970s their number peaked at 4.6 million, with cities in the Northeast representing the main attraction pole (63 per cent). In the 1980's the rural Northeast saw

Table 9.9

São Francisco River Basin: Poverty Index by State (Northeast only)

State	*No. of Municipalities within Basin*	*No. of Municipalities with Per cent of Poor Families*			
		Below NA	*NA to 40%*	*40 to 50%*	*Over 50%*
Goiás	3	0	3	0	0
Bahia	114	1	2	46	65
Pernambuco	65	0	5	29	31
Sergipe	27	0	1	24	2
Alagoas	49	0	1	23	25

Source: *http://www.codevasf.gov.br/vale/asp_sociais.htm*

34. The Gini coefficient grew from 0.59 at the end of the 1960's to 0.64 at the end of the 1980's. Rural coefficients were lower but grew faster in the period, from 0.47 to 0.54, while urban coefficients went from 0.60 to 0.64.

a net reduction in rural population. In 1990, poor rural families in the Northeast presented an average of 4.3 dependents, only 65 per cent of children aged 0-14 were enrolled in school, 66 per cent of family heads had no education, and less than 4 per cent had 4 or more years of education, 83 per cent reported subsistence farming as their primary occupation, from which families derived 76 per cent of their income. While recent analyses show a reduction in income inequality at the national level,[35] they largely confirmed the general results regarding poverty in the Northeast.

Macroeconomic policies and the sectoral nature of other public interventions contributed to widening intra—and inter-regional gaps in the Northeast, creating pockets of wealth in a still largely poor region, and determining the swelling of peri-urban areas by poor migrants from rural areas, particularly during droughts. The lack of integrated policies and investment programmes appears to have had a dampening effect on the benefits from the large number of small and large reservoirs in the region. When inadequate thought was given to the existing policies and interactions with other sector interventions, their capacity to induce complementary developments in the region was limited, often preventing expected impacts from actually materialising. In particular, the type of industrial development that resulted from macroeconomic and industrial policies may have had a dampening effect on their indirect impacts from inter-sectoral linkages, particularly those linked with the increased supply of electric power to industrial sectors. The next section estimates some of the indirect impacts of the studied dams.

Multiplier Analysis

Defining the Region's Boundary

The area of influence of the cascade of large reservoirs along the Sub-Médio São Francisco goes well beyond the immediately surrounding region. Their hydropower output represents the quasi-totality of energy use in the Northeast region. Their irrigation benefits have a more

---

35. A recent study (Ferreira *et al.*, 2001) of various income surveys available for Brazil revealed that data appear to systematically underestimate non-labour incomes, particularly for self-employed earners in rural areas.

localised impact, but have over time affected output composition, input use and factor remuneration in agriculture across the region, thanks to the dynamism of the irrigated region. If irrigation were the primary direct output of the dams, their analysis would need to focus solely on the São Francisco basin, or even on the relevant irrigated regions. Given the preeminence of hydropower output, however, the dams are likely to affect output, incomes and consumption levels for the whole Northeastern region. On the other hand, their impacts on the Brazilian economy as a whole, particularly their indirect and induced impacts, are likely to be limited, given the dependence of the Northeast's economy on demand and output from other regions and, conversely, the relatively small inter-industry impacts on the rest of Brazil of changes originating within the region (Guilhoto *et al.*, 2000). While data available for this study does not enable us to probe such conclusions, it seems reasonable to limit the focus of the analysis to the Northeast.[36]

Choosing the Right Analytical Tool[37]...and Using it Wisely

The choice of analytical tool should be driven by the assumptions regarding the mechanisms through which impacts are transmitted in the region of interest, with particular emphasis on the assumptions regarding factor employment and mobility. In demand-driven, fixed-price models, a change in demand for a sector's output results in changes in that sector's supply and factor demand. In the absence of supply constraints, adjustments occur via impacts on labour or capital employment, and via inter-regional factor migration. The presence of idle labour or capacity somewhere in the system—either locally or in other regions—is thus crucial for the existence of quantity-driven multiplier impacts as estimated by fixed price models. Existence of a

36. There is at least one outcome of the dams that may have significant repercussions outside the region, however. The impact of dams on labour demand—and thus on migration—are of particular interest for a region such as Brazil's Northeast, which has traditionally served as a source of migrant labourers for the rest of the country. Unfortunately, data and model limitations prevent a thorough treatment of the subject at this stage.

37. A number of analytical tools have been suggested in the literature for estimation of multiplier effects. See, for instance, Bell and Devrajan (1980), Hazell and Ramasamy (1991), Haggblade *et al.* (1991). See Bhatia *et al.* (2003) for a review of the literature and for a discussion of the strengths and weaknesses of each technique under different circumstances

multiplier depends on drawing unused or underused resources into more productive economic activities.[38] The prevailing conditions in the Brazilian economy in the period of our analysis seem to suggest this to be an acceptable assumption, at least for labour. For capital, however, this is probably too strong an assumption, given that it would assume a zero opportunity cost for the resource in a country and at a time where capital was scarce and return on investment other than the studied dams probably high, even if one accounts for the distortions engendered by the then prevalent import-substitution policies.

Despite this limitation, the analysis uses a fixed price Semi-Input-Output (S-I/O) model that lets all adjustments occur through quantity changes. The note regarding capital factors should be considered when building the model and interpreting its results, as its consequence is that indirect and induced impacts will likely overestimate the actual impacts that can be obtained in the presence of competition for capital from other sectors and regions. This was partially addressed by defining two models, one with no supply constraints and one with supply constraints on a number of capital-intensive sectors.

The model is part of the broad set of input-output (I/O) models. I/O analysis[39] traces the flow of production among the sectors of the economy, through to final domestic or export demand. Stemming from the I/O model pioneered by Leontief (Leontief, 1970; 1953), the analysis captures the interrelatedness of production arising through the flows of intermediate goods. Inter-industry linkages are based on a fixed coefficient, square Leontief matrix, assuming no substitution possibilities for producers. Similarly, factor remunerations represent fixed value-added proportions of total gross output. Input-output models evaluate indirect and induced economic impacts by computing multipliers that reflect the impact of changes in one sector's output—due to exogenous changes in final demand– onto the output of other sectors, income, and employment. It is important to note that, in this class of models, income is intended as total factor remuneration, i.e. value-added. The focus is on assessing the full impact of a project on factor income, and therefore on the income of households who own these factors. In input-

38. Haggblade *et al.* (1991).

39. For an in-depth discussion of I/O models see Hewings (1985).

output models, however, the lack of an explicit link between factor income and their distribution across household categories prevents the interpretation of value-added as a meaningful welfare measure.[40]

Semi-input-output (S-I/O) models are a variant of I/O models whereby a distinction is made between tradable and non-tradable goods. The former are assumed to have an exogenously set total output and in its portion sold domestically (or regionally), so that any demand shock will affect only their exports. In terms of domestic production, therefore, the whole brunt of demand shocks are borne by non-tradables. The implication of this distinction is that induced impacts—i.e. those descending from impacts on value-added/expenditure linkages—reverberate throughout the economy only via adjustments in non-tradables' output and their inter-industry linkages. This characterisation is important, as it refines the representation of the regional structure of production, reducing the risk of overestimating induced impacts when these are not felt by regional sectors or when tradables have their output determined by existing supply constraints.

I/O and S-I/O models can be used to compare output, value-added, and employment under the 'with dam' and 'without dam' scenarios.[41] For each scenario, the model computes their percentage changes in the 'with' *versus* 'without' cases, as well as summary measures of indirect to direct impacts, i.e. multipliers, defined as the ratio of the variation of total to direct impacts in the 'with' *versus* 'without project' scenario. Type I multipliers define total impacts as direct impacts plus the indirect impacts deriving from inter-industry linkages, while Type II multipliers also include impacts from induced income-expenditure

---

40. The author also consulted a 1995 Brazil national SAM (Cattaneo, 2002), whose agriculture accounts (activities, commodities, factors) are disaggregated by macro-region, including the Northeast. This SAM presents a meaningful disaggregation with respect to agricultural activities and commodities, and manufacturing activities include processed foods. Factors are also disaggregated by region, including skilled and unskilled urban and rural labour, as well as small and large farm capital. Households, however, are not disaggregated by region, preventing the use of this SAM to analyse distributional impacts in the Northeast.

41. Output (or Leontief) multipliers only reflect the degree to which industrial sectors are linked with each other and the strength of such linkages, but tell us nothing about the set of impacts on an economy of increased demand for the output of any of those sectors. To do so, value-added and employment multipliers are also estimated.

linkages. Care should be taken in their interpretation. Multipliers should not be read as indices of a project's desirability. Indeed, a small multiplier may be associated with a highly beneficial project that generates primarily direct impacts. On the other hand, large multipliers may translate into large overall net benefits depending on the initial importance of the directly impacted sectors in regional GDP and employment. However, the presence of large multipliers indicates that assessing the project based solely on its direct impacts may be missing a sizeable portion of its benefits.

The design and application of modeling techniques that compute multipliers can be useful for two reasons: (i) to highlight the salient paths of benefit generation associated with a project, and (ii) to clarify the interactions between a project's outputs and the structure of the economy and institutional/policy setting in which it is inserted. Instruments such as S-I/O models[42] can be used to explore the impacts on expected project outcomes of alternative assumptions about sector characteristics or policy packages. The use of these tools to analyse project outcomes in second-best conditions is particularly relevant, given the importance of understanding the potential outcomes of existing policy distortions.[43] Economy-wide models prove especially useful when distortions that may affect project outcomes are associated with policies whose impacts are not traditionally quantified in cost-benefit analyses (CBA), such as macroeconomic policies.[44] For large infrastructure projects whose impacts may affect the productive structure of a region and, conversely, can be affected by exogenous shocks and policy changes, this type of analysis can be a useful complement to traditional CBA. In addition, these techniques can be applied to estimate region-specific shadow prices, which would be needed for appropriate CBA but seldom exist (Bell and Devarajan: 1982).

42. And, preferably, price-endogenous models such as CGE models, which should be developed for the dams along the Sub-Médio São Francisco.

43. Again, price-endogenous models are more powerful in this respect, given the impact on prices of most policy-induced distortions, which cannot be captured by fixed price models such as S-I/O models.

44. Policies or aspects of the economy in which a project is inserted, are generally discussed in project assessment documents. However, their consideration is limited to the discussion of the risks affecting project benefits.

Therefore, the analysis should not focus solely on the numerical values of multipliers. What matters is the very process of building and analysing the regional database that embodies transactions in the regional economy, and the process of building the model and scenarios to be tested with it. This process may help identify not only the channels through which the majority of benefits and costs are produced (direct *versus* indirect *versus* income-induced), but also potential additional outputs or uses thereof—as well as unforeseen trade-offs and negative impacts.

The Model

A S-I/O model was calibrated for the Northeast region, with technical coefficients, shares and rates computed based on the 1992 regional I/O matrix. The model's key relationships are as follows (computed parameters in lower case, variables in upper case):

Production

A fixed-coefficients, Leontieff production function is assumed. Inputs include domestic and imported intermediates, labour and capital. The value of output is a linear combination of the values of all intermediate and value-added paid to primary factors. As output is valued at basic prices, which include indirect production taxes/subsidies, these must be included. Thus:

$X_j = \Sigma_i X_{ij} + \Sigma_{im} M_{imj} + \Sigma_{pr} V_{prj} + \Sigma_{it} IT_{itj}$; whose components are defined as follows:

Demand for Domestic Intermediates

$X_{ij} = a_{ij} * X_j$; for all i=1,2,...30 , j=1,2,...30 sectors

Value of intermediates purchased by sector *j* from sector *i*

= Technical coefficient * Value of gross output of sector *j*

Demand for Imported Intermediates

$M_{im,j} = b_{im,j} * Xj$; for all *im* = 1..3 , j = 1...30

Value of imported intermediates purchased by sector *j*

= Import Coefficient for import category *im* in sector *j* *Value of gross output of sector *j*

Demand for Primary Factors

$V_{pr,j} = p_{pr,j} X_j$; for all $pr = 1,2$, $j = 1...30$

Value-added of primary factor demand by sector j

= value-added share for factor *pr* in sector *j* *Value of gross output of sector *j*

Payments/Receipts of Production Taxes/Subsidies

$IT_{it,j} = t_{it,j} X_j$; for all $it = 1,2$, $j = 1...30$

Value of production taxes/subsidies paid/received by sector *j*

= tax/subsidy rates for sector *j* *Value of gross output of sector *j*

Regional Value-added (Total Factor Income)

Regional value-added is the sum of payments from production activities to primary factors.

$$TVA = \sum_{pr}\sum_{j} V_{prj}$$

Total Expenditure

In a model based on a I/O table, budget constraints are not imposed on each demand category. As opposed to SAM-based models, the factor income-demand account—expenditure loop is not closed in this class of models. However, it is required that— systemically—all resources be allocated. No explicit saving takes place in the model, which is short-run in nature. Therefore,

$$\Sigma_d EXP_d = TVA + \Sigma_{it} \Sigma_j IT_{itj}$$

where total expenditure for each demand category d (EXP_d) includes final demand for both domestic and imported goods. Note that net indirect taxes must be added, as overall expenditure includes the government account, which receives them and then spends them. In the 1992 Northeast I/O table, the government account does not appear separately, but rather is spread across the remaining final demand categories.

Final Demand

For each demand category, consumption shares by activity is assumed fixed at the initial average propensities. Thus:

$$FD_{i,d} = d\_coef_{i,d} * EXP_{d}; \text{ for all } i=1...30,\ d = 1...6$$

Total Demand for Output

Market equilibrium is ensured by requiring that all output be demanded, either in the form of intermediate inputs or as final products by final demand categories.

$$X_i = \Sigma_j X_{ij} + \Sigma_d FD_{i,d}$$

In a Semi-I/O model, output of tradables is fixed. It can be shocked exogenously in the scenarios, but cannot change endogenously in the adjustment towards a new equilibrium. In addition, domestic final demands for tradables are fixed, so that any adjustment to an exogenous increase in tradable output will be absorbed by changes in its exports. On the other hand, exports of non-tradables are fixed, and two closures are possible for their domestic final demands. The first fixes all domestic final demands. All adjustments in this case take place only via inter-industry linkages, so that the resulting multipliers are of Type I. A second option is to assume that changes in value-added only affect household consumption (or shared by all domestic demand categories, based on their initial shares of total expenditure), fixing all other final demand for non-tradables in the short run. This option captures both indirect and income-induced impacts, so that the resulting multipliers are of Type II. As no saving function is explicitly modeled, induced impacts are likely to overestimate actual impacts. In addition to the savings problem, as mentioned above, overestimation is inevitable due to the fixed-price nature of the model.

One way to reduce the risk of overestimating indirect impacts due to the idle capacity assumption is to choose tradables carefully. Their selection should depend on what is believed regarding the existence of idle capacity in the system and the impact the dam has on it. One option is to consider as tradables only the sectors that are directly affected by the dam, setting their output exogenously, estimating their direct impacts outside the model and then using them to shock the system. If a larger group of activities are expected to operate at capacity with respect to inputs other than water or electricity, they too may be treated as tradables, and their output and value-added will not be affected by the increased availability of the dam's output. The larger the number of supply-constrained sectors, the smaller the indirect impacts, induced impacts, and multipliers. Given the importance of electric power for the development of the industrial complex in the region, and given the role of the constraining factor played by water in agriculture, one may assume that the construction of one or more large dams in the Northeast would relax supply constraints sufficiently to treat all sectors other than those directly affected by the dams as non-tradables. On the other hand, given the number of model characteristics that may induce an overestimation of indirect impacts, a second version of the model is built with supply constraints on 12 capital-intensive industries, which are treated as tradables. The supply-constrained sectors[45] were chosen on the basis of their shares of exports in total demand, and based on considerations of capital availability in the Northeast. Thus, despite their high export to output ratios, livestock and food processing were excluded from the final set, which includes primarily capital-intensive industries. The supply-constrained version of the base drop the assumption of perfectly elastic capital availability. On the other hand, given that large volumes of public investments in capital-intensive sectors did in fact take place at the time of the construction of the dams, the scenarios described in the following paras are also run for the unconstrained base, so as to provide a range of possible impacts under alternative assumptions regarding capital availability.

The Database: 1992 Regional Input-Output Table

The model is calibrated based on the 1992 I/O tables for the Northeast macro-region.[46] These comprise 39 commodities (products), 30 activities (processes), a dummy account, and five final demand accounts (households, fixed investment, change in stock, export to the rest of

45. AAGR, AEXMIN, ASID, AMECH, AELETR, ATRAMAT, APAPULP, ACHEM, ATEXT, ALEATH, AOTH, ASIUP—see Appendix A.9.1 for a description of these sectors.

46. Consider *et al.* (1998). Their 1985 I/O table was also consulted for the analysis.

Table 9.10

Sector Participation in Output, Value-Added and Employment (Per cent)

	Output	*Value-Added*	*Employment*
Crop cultivation	8.15	9.28	31.77
Livestock production	3.51	4.61	13.61
TOTAL AGRICULTURE	11.66	13.88	45.38
Mineral extraction and manufacture (non-metal)	2.17	2.10	0.85
Steel production, Machinery, Vehicles	4.24	1.92	0.80
Wood & furniture, Pulp & paper, Printing	0.74	0.50	0.46
Chemical industry	7.98	3.91	0.63
Textiles, Clothing, Leather	4.34	2.36	3.43
Food Processing	6.29	3.41	0.98
Drinks	0.72	0.40	0.19
Miscellaneous manufacturing	0.26	0.21	0.22
Public Utility Services—Electricity, Gas, Water	5.01	4.30	0.46
Construction	8.65	8.21	5.10
TOTAL INDUSTRY	40.39	27.33	13.12
Commercial activities	9.75	10.28	13.52
Transportation services	3.88	2.84	2.68
Communication services	0.91	1.19	0.22
Finance and insurance services	7.92	11.92	0.81
Other public services	25.49	32.55	24.27
TOTAL SERVICES	47.96	58.78	41.49

Brazil, exports to the rest of the world).[47] Activities and final demand categories buy commodities and imports from the rest of Brazil or rest of the World, and pay commodity taxes on both purchases. Activities then pay wages/rents to primary factors and indirect taxes net of production subsidies.

Table 9.10 gives an idea of the structure of the regional economy, displaying the shares of output, value-added and labour demand[48] of the major sectors or clusters of sectors presented in the more disaggregated table in Appendix A 9.1. A couple of characteristics can be highlighted. As was discussed earlier, agriculture represents a declining share of GDP

47. For ease of use in the model, the full set of tables in the 1992 I/O table was used to derive a 30x30 activities to activities table. The 30x30 I/O table is presented in Appendix A 9.1, together with a legend spelling out the names of each account. The full set of 1992 tables is listed in Appendix A.9.2.

48. Based on employment data that is not reported in Appendix A.9.1, but contained in the tables described in Appendix A.9.2.

but remains the largest employer in the region. Food processing, on the other hand, represents a sizable share of the region's GDP compared with other powerful clusters, such as the chemical industrial complex, highlighting the importance of the agro-industrial productive chain for the region's economy.

Before defining the scenarios to be simulated, the unconstrained I/O value-added multipliers were computed for all sectors, some of which are presented in Table 9.11. The reason was to get a feel for the magnitude of unconstrained[49] indirect and induced impacts of different sectors, and show how information that could be rapidly extracted from the I/O database can be used to build scenarios by complementing observations from the qualitative analysis and literature review.

Table 9.11

Unconstrained Value-Added Multipliers in Key Sectors[50]

Economic Sector	*Type I*	*Type II*
Crop Production	1.263	3.010
Food Processing	2.389	5.694
Drinks	2.145	5.114
Electricity and Water	1.511	3.602

Large multipliers on food processing and drinks confirm the observations on the importance of linkages between irrigated agricultural production and agro-processing activities, as these have the potential to spur value-added in areas where irrigated agriculture is present.[51] Unfortunately, the 1992 I/O table only contains one aggregate crop

49. Note that here 'unconstrained' means that no sector is treated as tradable, not even the one whose impacts are being computed, so that extra indirect and induced impacts can reverberate through this sector as well. Hence, these multipliers are larger than the 'unconstrained' multipliers described in the following sections.

50. These multipliers are obviously larger than those found based on the constrained S-I/O model.

51. Studies carried out in the context of the Banco Nacional do Nordeste/Inter-American Development Bank program *New Irrigation Model for the Northeast* call for a broader approach to agriculture production in the region, to consider the full productive chain and its impacts on on- and off-farm sources of income for local populations. Accarini (2002); Haugenauer *et al*. (2001), Prochnik and Hauguenauer (2001).

cultivation sector. This is problematic for the analysis, as the expansion of irrigation in the region did not only affect planted area and yields, but rather encouraged a profound restructuring of the agriculture sector in the affected areas. Changes in cropping patterns affect the value of output and value-added of the region, but more importantly spur a whole new set of backward and forward linkages with agro-processing and other activities. The impacts of irrigation on activities in the agribusiness productive chain cannot therefore be confined to the backward linkages and expenditure impacts produced by a direct impact on crop output, which is the only type of analysis permitted by the 1992 I/O. The model's estimates of the impacts of irrigation, based solely on output increases in agriculture, are thus likely to be lower end values of its overall impacts.

The Scenarios

A traditional intra-regional, single-period analysis is applied to the 1992 I/O table. The model studies the impacts of eliminating one or more of the studied dams on the 1992 economy—a year of normal operation for them.[52] The analysis does not attempt to produce a full counterfactual describing alternative development paths the region might have followed in their absence. When comparing 'with' and 'without project' scenarios, the structure of the region's economy remains that of the 1992 I/O table, i.e. the structure that developed 'with' the dams in place. What is eliminated in the 'without project' case are only the outputs directly linked with the project, and their indirect and induced impacts. The goal of this exercise is to better characterise the nature and relative magnitude of direct, indirect and induced benefits derived from the dams in a normal year of operation, not to try to derive a full alternative development path for the region.

The basic set of scenarios combine alternative assumptions regarding changes in hydropower generation and crop output if one or more of the reservoirs on the Sub-Médio São Francisco were to disappear

52. It is important to note that while technically this may be true, as maturity had been reached, 1992 appears far from being an ordinary year for the region. A drought was under way and the hyperinflation period was reaching its critical peak. The impossibility to obtain a more recent I/O table, however, made it impossible to obviate such problems. Note that a number of recent economic studies of the region have also been based on the 1992 I/O.

in 1992. To test the impacts of changes in the productive structure of regional agriculture due to irrigation availability, a second set of scenarios suggests ways to include an explicit characterisation of the impacts on the food processing sector. The scenarios are run under two alternative closures for non-tradables, i.e., with 'endogenous' final demand by households (including induced impacts), or with 'exogenous' final demand (inter-industry linkages only), generating Type II and I multipliers, respectively. The experiment results are all also run starting from the second base model with a higher number of supply-constrained, tradable sectors, to check the range of variation of indirect impact and multiplier values.

Single-Sector Scenarios

Power Generation Assumptions

CHESF owns and operates the hydropower system along the Sub-Médio and Baixo São Francisco. Overall CHESF accounts for over 80 per cent of the region's installed capacity, and about 90 per cent of its power generation.[53] In 1992, the hydropower complex in the Sub-Médio region—excluding Xingó—accounted for 90 per cent of CHESF's installed capacity, other hydropower for 4 per cent, and thermal generation for 6 per cent. Specifically, Sobradinho accounted for 14 per cent of installed capacity, PA I-III for 18.8 per cent, PA IV for 32.65 per cent, Luis Gonzaga for 19.6 per cent, and Apolônio Sales for 5.3 per cent.

HLO: Sobradinho is not built, with no impacts on the output the downstream stations existing in 1992 (i.e. they are not negatively affected by less reliable flows, nor do they increase their share in generation). No additional thermo or hydropower generation is installed. Power generation declines

53. Data from *www.chesf.com.br*, *www.energiabrasil.gov.br* and *www.eletrobras.gov.br.* Data about CHESF found on the latter state that in 1999 it represented only 72.1 per cent of the region's generation. However, a comparison of its data with those found at *www.energiabrasil.gov.br* contradicts this share in favour of a 90 per cent share. The ELETROBRAS document sets CHESF's commercialised energy at 45,563 GWh. Subtracting from this the energy imported from the North (4,635 GWh according to ELETROBRAS, 4,960 GWh according to *energiabrasil*) results in a share of over 90 per cent of the sum of energy generated by hydropower and conventional thermal power (44,676 and 210.g GWh, respectively) reported by the *energiabrasil* site for the Northeast region.

by 14 per cent and the output of activity ASIUP decline by 0.14 * 0.90 * 0.85 (=11 per cent), where 0.9 is CHESF's share of regional power output, and 0.85 is the share of ASIUP output value pertaining to electricity, according to the 1985 I/O table.

HME: Sobradinho and Itaparica are not built, and we assume that this would force Paulo Afonso IV to operate at 50 per cent of 1992 capacity. We also assume that both small hydropower developments and thermal capacity would increase 2.5 times. Power generation declines by 43 per cent and the exogenous change in ASIUP output is 0.43 * 0.90 * 0.85 (= 33 per cent)

HHI: the extreme scenario assumes that Sobradinho and Itaparica dams are not built, that this prevents the operation of PA IV, and that no additional small hydro or thermal capacity is added to the system. The worst case scenario sees a 66 per cent decline in power generation, equivalent to a 50.5 per cent decline in ASIUP output.

Crop Production Assumptions

The flow regulation function played by large dams along the São Francisco was crucial in making large-scale irrigation viable in the Sub-Médio São Francisco. To compute the impact that a 'without project' reduction in irrigated areas would have on the value of agricultural output in the Northeast region, the following formula was used:

Per cent change in output value =

$$\frac{(Ar^{WO} * Yr * Pr + Ai^{WO} * Yi * Pi) - (Ar^{W} * Yr * Pr + Ai^{W} * Yi * Pi)}{(Ar^{W} * Yr * Pr + Ai^{W} * Yi * Pi)}$$

Where: Ar^{WO} and Ar^{W} are rainfed area without and with project, respectively, and $Ai^{WO}Ai^{W}$ stand for irrigated area, *Yr* and *Yi* are rainfed and irrigated yields, and *Pr* and *Pi* are the average prices of the crop mix produced by rainfed and irrigated farming. Expressing yields and prices relative to rainfed yields and prices, the expression can be simplified as follows:

Per cent change in output value =

$$\frac{\left(Ar^{WO} + Ai^{WO} * \frac{Yi}{Yr} * \frac{Pi}{Pr}\right) - \left(Ar^{W} + Ai^{W} * \frac{Yi}{Yr} * \frac{Pi}{Pr}\right)}{\left(Ar^{W} + Ai^{W} * \frac{Yi}{Yr} * \frac{Pi}{Pr}\right)}$$

The set of data available for the region as a whole that are closer to the 1992 benchmark year are those for 1995-96[54], when the Northeast's area under perennial and annual crops was 10,354,388 ha, of which 400,000 ha were irrigated. These are the relevant 'with project' area data. To compute the 'without project' area, we also use 1995 data for the region. Table 9.3 shows that 1995 irrigated area in the São Francisco basin was 300,000 ha, while Table 9.4 shows that in recent years the irrigated area in the large dams sub-region represents about 47 per cent of the basin's irrigated area. Assuming this share was valid in 1995 as well, the resulting 1995 irrigated area in the dam's sub-region is about 140,000 ha. One final assumption is that the 29 per cent ratio of public to total irrigated areas prevalent in 1990-1994 for the basin as a whole (Table 9.3) is also valid for the large dams region, resulting in a public irrigation area of approximately 41,000 ha.

To estimate the impact of shifting part of this from irrigated to rainfed farming, we also need to make assumptions regarding differences in the value of output for irrigated *versus* rainfed farming. As was discussed earlier, irrigation is associated with yields that are on average 1.5-2 times those obtained on rainfed land, and it was estimated that farmers' rents in irrigated agriculture can be much higher and more stable than those in rainfed agriculture, due to a combination of the higher value of crops grown in irrigated areas and the reduced uncertainty of water availability. In particular, the irrigated perimeters developed by CODEVASF along the Sub-Médio São Francisco have shown great success in switching to fruit production, which has been shown to produce net revenues that are up to five times higher than those obtained by other irrigated crops (Sampaio, 1994). Overall, FAO

54. Agricultural census, IBGE, (1996). Data can be downloaded from: *www.sidra.ibge.gov.br.*

1995 price data for Brazil shows that the average price of the crop mix in irrigated agriculture is twice that obtainable in rainfed agriculture (*www.fao.org*). For the present analysis, the following three sets of assumptions were made:

ALO: Only public irrigation projects in the Sub-Médio São Francisco—downstream of Sobradinho and upstream of Xingó—are not undertaken if the dams are not built, resulting in a 41,000 ha reduction in irrigated area. This land is assumed to be put to rainfed production, which is assumed to produce yields and prices of the crop mix that are 50 per cent of those for irrigated agriculture. Under the without project scenario, we would thus witness a 1.1 per cent decline in the value of crop production (AAGR). Livestock sector output (APEC), on the other hand, is left unchanged, as the impacts of irrigation expansion on the sector are not clear. While increases in rents and crop yields may lead to increases in productivity via the availability of better feeds and other productivity-enhancing measures, one may expect farmers to move out of traditional livestock activities in favour of crop production. The net impact on livestock output is not clear and will therefore, be ignored.

AME: All irrigated land in the Sub-Médio São Francisco area (141,000 ha) becomes rainfed. The same assumptions regarding yield and rent differences hold as under ALO. Under the without project scenario, AAGR output value declines by 3.7 per cent.

AHI: The extreme value adopted for the without project case is a 75 per cent decline in the 'basin's' irrigated area (225,000 ha). The same assumptions regarding yield and rent differences hold as under ALO. This results in a 5.85 per cent decline in AAGR output value

Combined Scenarios

Combining these sets of assumptions, we obtain a grid of nine combined without project scenarios (Table 9.12).

Table 9.12

The Nine Combined Without Project Scenarios

	HLO	*HME*	*HHI*
ALO	–11% AIUP output value	–33% AIUP output value	–50.5% AIUP output
	–1.1% AAGR output value	–1.1% AAGR output value	–1.1% AAGR output value
AME	–11% AIUP output value	–33% AIUP output value	–50.5% AIUP output value
	–3.7 % AAGR output value	–3.7 % AAGR output value	–3.7 % AAGR output value
AHI	–11% AIUP output value	–33% AIUP output value	–50.5% AIUP output value
	–5.85% AAGR output value	–5.85% AAGR output value	–5.85% AAGR output value

The Results

Single-sector Scenarios

When each directly affected sector is considered in isolation and the three scenarios defined for each are run, multipliers do not change across scenarios, but do change when the set of supply-constrained (tradable) activities is changed. On the other hand, percentage changes in value-added, output and employment obviously change across scenarios, and are also affected by changes in the tradables set.

Table 9.13 presents the multipliers and percentage changes for value-added and employment, assuming that no sectors, apart from the directly affected one, are supply-constrained. Percentage changes (%D) should be read as increases in total value-added and employment generated in the 'with project' (i.e., the status quo) *versus* the 'without project' case. They too are distinguished between those including induced impacts (II) and those including direct impacts and indirect impacts from inter-industry linkages only (I)

The weight of income-expenditure induced impacts on value-added changes is large in both single-sector scenarios, representing 50-54 per cent of total value-added effects. With respect to employment, however, the two scenarios behave quite differently, with induced impacts representing over 85 per cent of employment creation in the hydropower scenarios *versus* 20-25 per cent for the agriculture scenario. In the former, their size calls for caution when interpreting results, which are likely to overestimate the actual impacts of single-sector output

Table 9.13

With versus Without Project Value Added and Employment Multipliers and Percentage Changes: Unconstrained Single-sector Scenarios

SECTORS	*M II VA*	*M I VA*	*M II EMPL*	*M I EMPL*
Hydropower	2.67	1.25	16.436	2.70
Agriculture	2.292	1.165	1.320	1.05
SCENARIOS	%DVA II	%DVA I	%D EMPL II	%D EMPL I
HLO	1.2%	0.6%	0.8%	0.1%
HME	3.9%	1.8%	2.6%	0.4%
HHI	6.1%	2.8%	4.0%	0.6%
ALO	0.2%	0.1%	0.5%	0.4%
AME	0.8%	0.4%	1.6%	1.2%
AHI	1.3%	0.6%	2.5%	2.0%

increases, due to the model's inability to account for saving rates, its fixed price nature, as well as the extreme assumption of no supply constraints on any activity besides the directly impacted one. This assumption can be relaxed, however, and all scenarios can be run with 12 supply-constrained activities. As shown in Table 9.14, the results that are affected the most are those linked with hydropower, as some of the capital-intensive industries that are now supply-constrained present significant linkages with the electricity sector. Table 9.15 shows by how much the unconstrained multipliers and percentage changes in value-added and employment exceed their supply-constrained counterparts. These differences are very high for income-expenditure (Type II) linkages, particularly for employment in the hydropower scenarios, but low when only inter-industry linkages are considered.

Thus, assumptions regarding supply constraints do matter, and can be partially captured with appropriate scenarios even with a fixed price model. Yet, even after correcting for supply constraints, income-induced impacts continue to dominate the hydropower-only scenario, at least with respect to employment generation, where they represent 80 per cent of overall indirect impacts.

Table 9.14

With versus Without Project Multipliers and Percentage Changes: Supply-Constrained Single-sector Scenarios

SECTORS	*M II VA*	*M I VA*	*M II EMPL*	*M I EMPL*
Hydropower	2.114	1.162	10.009	2.185
Agriculture	1.915	1.103	1.253	1.043
SCENARIOS	%DVA II	%DVA I	%D EMPL II	%D EMPL I
HLO	1.0%	0.5%	0.5%	0.1%
HME	3.1%	1.7%	1.6%	0.3%
HHI	4.8%	2.6%	2.4%	0.5%
ALO	0.2%	0.1%	0.4%	0.4%
AME	0.7%	0.4%	1.5%	1.2%
AHI	1.1%	0.6%	2.4%	2.0%

Table 9.15

Unconstrained versus Supply-Constrained Single-sector Scenarios

SECTORS	*M II VA*	*M I VA*	*M II EMPL*	*M I EMPL*
Hydropower	26.3%	7.6%	64.2%	23.6%
Agriculture	19.7%	5.6%	5.35%	0.7%
SCENARIOS	%DVA II	%DVA I	%D EMPL II	%D EMPL I
HLO	20.0%	20.0%	60.0%	–
HME	25.8%	5.9%	62.5%	33.3%
HHI	27.1%	7.7%	66.7%	20.0%
ALO	–	–	25.0%	–
AME	14.3%	–	6.7%	–
AHI	18.2%	–	4.2%	–

Combined Scenarios

Table 9.16 presents the multipliers associated with the nine combined, two-sector scenarios, run using the supply-constrained assumptions.[55] The combined scenarios produce multipliers that fall between the

55. Under the unconstrained supply assumption, where only AAGR and ASIUP are treated as tradables, multipliers are 10-18 per cent larger. VA Type II multipliers range from 2.28 to 2.4 and VA M I from 1.19 to 1.24. Employment Type II multipliers, on the other hand, range from 1.6 to 5.6, and EMPL M I from 1.09 to 1.64. As in the supply-constrained case, the lowest values correspond to the HLO/AHI case, while the highest to the HHI/ALO scenario.

Table 9.16

Multipliers under Supply-Constrained Combined Scenarios

SCENARIOS	*M II VA*	*M I VA*	*M II EMPL*	*M I EMPL*
HLO + ALO	2.078	1.151	2.344	1.185
HLO + AME	2.029	1.136	1.609	1.089
HLO + AHI	2.007	1.130	1.482	1.073
HME + ALO	2.101	1.159	3.916	1.390
HME + AME	2.075	1.150	2.260	1.174
HME + AHI	2.059	1.145	1.918	1.130
HHI + ALO	2.105	1.159	4.769	1.501
HHI + AME	2.087	1.154	2.709	1.233
HHI + AHI	2.074	1.150	2.234	1.171

multiplier values of the single-sector scenarios. This is because multipliers represent average measures of the relative importance of indirect (or indirect + induced) to direct impacts, so that the peaks reached by the induced impacts of hydropower output changes—particularly with respect to employment creation—are softened by its interaction with the more direct impact-heavy agricultural sector. Thus, under most scenarios, induced impacts account for 28-50 per cent of employment multiplier impacts, climbing over 60 per cent for only two scenarios (HHI/ALO and HME/ALO—not surprisingly, scenarios where low agricultural output changes interact with medium-high hydropower output shocks).

As in the single-sector scenarios, Type II multipliers are quite large and driven by income-expenditure induced impacts, while inter-industry linkages do not produce very significant indirect impacts. Inter-industry linkages produce only a maximum of 16 extra centavos for each Cr$ of direct value-added impact, while induced impacts produce an additional 81-92 centavos per CR$ of direct value-added impact, accounting for about 45 per cent of the Type II income multipliers. On the other hand, value-added M II close to 2.0 are largely driven by the significant impacts on service sectors. With respect to employment creation, this is concentrated mainly in the commercial and public service sectors, which are absorbing a growing share of the region's workforce. Value-added and employment impacts reported in Table 9.17 show that the outcomes of

Table 9.17

Percentage Changes With versus Without Project: Supply-Constrained Scenarios

SCENARIOS	%DVA II	%DVA I	%D EMPL II	%D EMPL I
HLO + ALO	1.2%	0.7%	0.9%	0.5%
HLO + AME	1.7%	0.9%	2.0%	1.4%
HLO + AHI	2.1%	1.1%	2.9%	2.1%
HME + ALO	3.3%	1.8%	2.0%	0.7%
HME + AME	3.8%	2.1%	3.1%	1.6%
HME + AHI	4.2%	2.3%	4.0%	2.3%
HHI + ALO	5.0%	2.7%	2.9%	0.9%
HHI + AME	5.5%	3.0%	4.0%	1.8%
HHI + AHI	6.0%	3.2%	4.9%	2.5%

combined scenarios are close to a simple sum of the impacts of single-sector scenarios.

Finally, on a sector by sector basis, the largest indirect and induced impacts are generated in services, particularly communications and other public services, in clothing and pharmaceuticals/fragrances, and in the production of drinks. When the latter is considered together with food processing, the indirect impacts, that agriculture and hydropower output changes generate in this complex, dwarf all remaining single sector impacts.

The results show that the dams may have accounted for 1.2 to 6 per cent of the Northeast GDP, depending on which 'without project' scenario is deemed more realistic. Value-added Type II multiplier values are close to 2.0 in most scenarios. For every unit of value generated by the sectors directly affected by the dams, another unit could be generated in the region. These secondary impacts are largely due to income-expenditure induced effects, however, as the effects of pure inter-industry linkages are much lower, with Type I multipliers around 1.15. This result is in line with what other authors have reported

56. Guilhoto (1998), Guilhoto *et al.* (2001), Haddad and Hewings (2000). Based on an inter-regional decomposition of inter-sectoral linkages, these analyses were able to point out that changes in the rest of Brazil had repercussions in the Northeast, while changes originating in the Northeast did not produce significant impacts on other regions, and therefore did not reverberate back onto the Northeast economy via their demands.

regarding the dependance of the productive structure of the Northeast region on demands from the rest of the country.[56] With respect to employment, the range of multipliers is larger than for value-added, showing that conclusions regarding employment creation by dams are sensitive to the choice of scenario. One may note that employment multipliers from agriculture are much lower than those from hydropower. This is partly a result of the higher labour-intensity of crop production compared to the electricity and water sector that hydropower is part of. Thus, while irrigation has a relatively much higher impact on total employment generation, this results primarily from direct impacts within the sector than from indirect employment generation. This appears to contradict some of the findings regarding induced employment creation by irrigation. However, one should remember what was said regarding one limitation of the model, namely its incapacity to fully capture the impacts of dams on the structural changes irrigation induces in crop production. Thus, these results are bound to underestimate the full set of indirect impacts produced by irrigation.

Tables 9.14 and 9.17 present an interesting picture of the impact of different outputs of dams on the region's economy. Relatively large hydropower output shocks produce larger overall value-added than simulated reductions in irrigated area, and yet these differences do not appear as large when one considers how small irrigated areas and their changes are compared to the region's planted area. This is true even when the supply constraints on some capital-intensive and energy-intensive sectors are removed. The results with respect to employment are even more striking, as the switch to irrigation of small areas produces a same percentage change in employment as large reductions in hydropower output, under the supply-constrained assumption. These results are consistent with other authors' findings regarding (i) increases in the share of regional GDP accounted for by dynamic irrigated areas *versus* the stagnating sertão, (ii) demand-induced impacts of droughts, and (iii) impacts of irrigation development on employment and value-added.[57]

57. See, for instance, Maia Gomes and Vergolino (1995), Bonelli (2001), Rodrigues (2001).

Beyond the magnitude of these impacts, what matters is the information they provide regarding the structure of benefits that can be attributed to the dams. The results suggest that larger overall impacts might have been achieved if more attention had been given to multi-purpose use of their water. The dams were built primarily to feed the hunger for electricity of the Northeast's industrial complex. Due largely to macroeconomic and industrial policies from the late 1960's to the mid-1980's, its excessive capital intensity and lack of significant inter-sectoral linkages may have impaired the dams' capacity to generate the larger impacts one would have expected. While the model shows the presence of indirect impacts, these descend primarily from income-expenditure effects on service demand, which may be overestimated due to the fixed price nature of the model. In the absence of the dams, alternative, more expensive sources of electricity would have been brought on board, which may have favoured a restructuring of the industrial complex in favour of less energy-intensive sectors. These might well have produced larger inter-sectoral linkages, in the end benefiting the region. It is unfortunate that the links between agriculture and agribusiness cannot be appropriately analysed with the help of this model, given the mentioned limitations of the 1992 I/O table. If this were possible, one could study whether larger indirect impacts could be associated with a reduction of irrigated areas.

Social and Environmental Impacts: An Overview[58]

Social and environmental impacts generated by the construction of the cascade of reservoirs, have been the object of serious scrutiny over the past decade. GDP growth associated with the dams' construction did not come without negative environmental and social externalities, and at least initially it had negative distributional impacts. Hydropower development benefited primarily the already favoured coastal metropolitan regions, where energy-intensive industries tended to concentrate. The recent successes attributable to the modernisation of agriculture, its link with agribusiness and the spillover effects from the interactions of small and large farmers followed years of peasant

58. For a more detailed analysis of resettlement issues and social impacts of dams in Brazil, see Ferradás (2001), on which this section draws heavily.

displacement and increasing land concentration. Resettled peasants were often given plots with poor soils or in areas that required agricultural techniques they were not familiar with. For peasants who became landless, employment in rural areas materialised slowly and only in small irrigated areas, while industries were generally in distant areas and many of them either required little labour—due to the bias in favour of capital-intensive industrialisation of ISI policies—or skilled labour for which they had no training. In addition, the irrigated area boom has not been reflected in a marked improvement in other areas of the Northeast, so that intra-regional imbalances probably worsened. The lesson learnt from these experiences is that appropriate parallel interventions need to be undertaken to minimise the social externalities of dams, and that these may at the same time improve their economic outcomes.

With regards to social impacts, the main issues to discuss relate to the resettlement linked to the construction of both Sobradinho and Itaparica, specifically:

a. Their forced nature and the secrecy and lack of consultation regarding resettlement and other project aspects.

b. The limited and often inappropriate options given to resettled populations, and the lack of programmes to support the adjustment of populations to new conditions .

c. Failure to recognize as affected population groups other than property owners in the flooded area, such as *vazanteiros*, fishermen, people employed by those residing in affected areas, etc.

d. Specific impacts on indigenous people

e. Negative impacts of migration to growing urban areas, such as the swelling of slums in and an increase in urban violence.

The inundation of areas upstream of Sobradinho forced the resettlement of the inhabitants of four cities (Casa Nova, Remanso, Sento Sé e Pilão Arcado) and 30 smaller centres, and of the population living in rural areas of the municipalities of Juazeiro and Xique-Xique. During construction, local opposition surged and the government used military intervention to evict people. In all, 11,853 families (over 70,000 people) had to abandon the area by the end of 1977. Of these, 5,806 families remained in rural plots around the Sobradinho lake, while the remaining population moved to the new location of the resettled municipalities. In addition, downstream flow regime changes also resulted in the permanent flooding of 32,000 ha of rice and the displacement of 50,000 persons (Gutman, 2001). The resettlement of the affected population was poorly planned and executed, and while resettlement costs escalated five-fold (resulting in overall costs overruns of 39 per cent), they failed to address the needs of the displaced population. Many of the new settlements failed to attract the target population. Displaced families received little or no compensation, as few could legally prove ownership of the abandoned plot, and even in this case, compensation did not always materialise. As a consequence, a large portion of the compensation went to the better off, and land concentration and poverty both grew. Most displaced families relied for subsistence on floodplain agriculture, but were instead relocated in dry areas where traditional agricultural practices were infeasible and where most of their livestock died. They were provided no technical assistance or other means to adjust to the new situation, and received no compensation for their losses. Most of the resettled villages lacked electricity, even though the dam was mainly built for power generation. There were no legal mechanisms through which the population could make claims or demand information, as secrecy prevailed under the military regime. Very little information was provided to local authorities and the general public, and access to construction sites and dam project headquarters was also very difficult. Security controls discouraged people's involvement. Access to basic information on social and environmental impacts is apparently still difficult. (Ferradás, 1998). Lessons were learnt from this tragedy, however, as the outcry that followed it contributed to the creation of the World Bank's *Resettlement Policy*.

A crucial problem with early resettlement and compensation was the inability to identify the full set of affected parties, as it was believed that only those living in submerged areas would be affected by dam construction. This overlooks the fact that the livelihood of many in the region was based on the seasonal use of diverse natural environments. Even people whose homes and cultivation plots are in areas to be flooded might derive part of their livelihood from activities depending on other areas that might be affected by deforestation in flooded areas, changes in water supply and/or water quality. Indigenous communities are even more vulnerable because very often their subsistence and socio-cultural ties go well beyond the site where they had built their homes (Ferradás, 2001). In addition, activities in the informal economy were not assessed for relocation purposes. In the early Brazilian dam projects, *vazanteiros* (river bank dwellers and occasional farmers relying on the natural flooding and retreating cycles of the river, located downstream of the dams and affected by their alteration of natural flows), fishermen and artisans were often overlooked, and the few who were compensated were often given homes in cities.

Another problem was the lack of voice many stakeholders had in early negotiations, which were generally limited to the energy company and some state agency acting for the affected population, often in an authoritarian and paternalistic manner. Decisions were mostly made without consultation to regional interests, and local governments generally had little knowledge of the characteristics of projects and their impacts. Their negotiating power was also very limited, due in part to the centralisation of military regimes. Since the mid-late 1980s the democratisation process enabled new actors to participate in the negotiation of dam projects, including political parties, non-governmental organisations, grassroots movements, academic and research institutions, and local governments. Non-governmental and grassroots organisations are particularly active and have led to significant changes in the process. However, not all these groups share the same visions regarding key problems and solutions, and the degree to which they represent affected people is sometimes a reason for concern. Today many previously excluded members of affected communities have organised and dam agencies are facing lawsuits and peoples' claims—for harvest losses, loss of economic activities, inadequate compensation, damages caused by changes in water quality or inappropriate relocation sites—showing that if social damages are not adequately compensated, it might be more costly to deal with them later.

---

59. The discussion of the Itaparica is largely derived from World Bank (1998), the 1999 Itaparica *Project Status Report*, discussions with Bank officials in 1997-2000, and a number of documents by Rede Brasil and other NGOs that were involved in the process, including Barros (2001).

An attempt to learn from Sobradinho's experience was the Itaparica Resettlement and Irrigation Project.[59] Partially funded by the World Bank, which did not finance the construction of the dam, it was the first example of stand-alone resettlement project, designed solely to benefit about 8,100 families displaced by the Itaparica Hydroelectric Project. The project was supposed to develop about 18,000 ha of new irrigated land for almost 6,000 families, including *vazanteiros* and other previously unrecognised affected groups.

The US$ 232 million project included construction of 110 'agrovilas', or agricultural settlements, and six irrigation projects for some 40,000 people. Despite its good intentions, it has over time encountered innumerable difficulties in achieving its objectives. Among these difficulties were inadequate counterpart funding leading to delays in irrigation systems installation; lack of agency coordination; the macroeconomic turmoil which hit Brazil in the early 1990s; and problems directly related with the design of the project, especially the assumption that unskilled farmers could successfully undertake irrigation with advanced technologies on soils that required sophisticated management techniques. Training activities were provided, but didn't initially prove effective. By the late 1990's, local communities were complaining that only one third of the promised projects had been built, that much of the irrigation infrastructure was falling apart because of poor construction and materials, and that irrigation schemes had proven too expensive to use. In addition, numerous problems have plagued access to its water, leading to allegations that this was being used to irrigate drug plantations instead of the smallholders' plots it was designed for. In 1997, these and other allegations of funding diversion and corruption, led communities' representatives to submit a claim to the Bank's Independent Inspection Panel on Dams regarding the possibility of sending an inspection mission to the project.[60] This was rejected by the Bank's Board of Directors, in favour of a commitment by the Brazilian government to solve Itaparica's problems with Bank supervision. Communities' representatives, however, state that they were given little access to the decision-making regarding the new funds'

60. In October 1997, one of the community representatives who had participated in drawing the claim was assassinated, possibly by drug traffickers.

allocation, and no role in their monitoring. This process was painful for all parties involved, leading the representatives of affected populations to question the Bank's willingness to assume responsibility for its mistakes, but also raising questions about the politicisation of the residents' association. Today, construction continues and the situation is improving, but the community still largely depends on direct monthly transfers (the 'monthly maintenance subsidy') provided to each family, and strong opposition has faced any attempt to phase out the monthly checks. The sad irony is that establishing and maintaining this broken communities cost more than US$190,000 per family. Itaparica appears to be an example of good intentions gone awry. When failure became evident, the new World Bank managers who took over a failing project had at least the courage to renew their commitment to the involved communities, and the project is showing some reason for hope. Since Brazil's financial crisis, however, some of the funds promised in 1997 were cut. The experience shows that more needs to be done to balance the interests and requests, and to recognise and address the shortcomings, of all parties involved, from affected populations, to the groups that represent them, to the company in charge of the project, to all levels of government.

A final concern is that local communities suffer the adverse impacts for project whose main benefits—those linked with hydropower generation—are appropriated by industries and urban area dwellers located in coastal areas, worsening the intra-regional income gap. Brazil addressed this issue, at least in part, by setting up a system of royalties to be paid to municipalities affected by dam construction as a compensation for lost development opportunities. The provision of energy generated by dams in the regions affected by them is also required.

Law 7990 (1989) establishes that states, federal districts and municipalities should be compensated for the use of their hydraulic resources for the production of energy. Law 8,001 (1990) defines how compensation should be distributed. Recently, the overhauling of Brazil's water resources management system has led to a revision of royalty payments, with shares of these going to the National Water Agency as payment for the water stored and used for generation.

Environmental externalities of the dams in the Sub-Médio São Francisco seem to have been less dramatic than the social impact. The major concerns are the impacts on riverine and estuary ecosystems, biodiversity and health from:

a. Reservoir construction and associated activities such as road construction.

b. Flooding of upstream areas.

c. Changes in natural flows.

d. Sediments from deforestation along river banks—an increasing problem, particularly for fish harvests.

e. Negative development impacts, such as the increased volumes of untreated municipal and industrial wastewater.

f. Lack of fish ladders for species spawning upstream and introduction of foreign species.

Particularly relevant are the adverse impacts on irrigated agriculture, as well as their potential mitigating measures, given current plans to substantially expand irrigation in the region. Modernisation, as well as the need to utilise less fertile soils, as large areas of more fertile soils closer to the riverbed were submerged, led to the increased use of fertilisers and pesticides. Their use has often been excessive, and so have been water applications, leading to agrochemical contamination as well as soil salinisation, which is worsened by the lack of adequate drainage in semi-arid irrigated areas. Addressing the rational use of both water and agrochemicals is possible, given their interactions, and crucial at a moment of mounting water use conflicts and when water diversions from the São Francisco river are being planned to support irrigated agriculture.

Early dam projects were generally undertaken in the absence of appropriate, comprehensive environmental impact assessments. Brazil now has legislation requiring environmental and social assessments before projects are approved.[61] Since 1991, the Plano Diretor de Meio Ambiente (PNMD), states that social planning in the energy sector

---

61. For a list of Brazil's Laws and Decrees regulating environmental, social, and indigenous issues, see Muller, 1995

should include: a) the identification of social impacts induced by the project, b) studies and propositions of options to mitigate the social and economic impacts, c) public consultation and revisions in the feasibility phase, d) presentation of alternatives to improve social and environmental conditions to the administrative sector (state institutions), in all the impacted areas, e) design and execution of development plans, unifying criteria, establishing agreements with various institutions to achieve social and economic regional sustainability, (Muller 1995). Resolutions by CONAMA (the National Environmental Council) no. 001/1986, 006/1987 and 10/1987 deal with Environmental assessments and licensing of large electric projects. Finally, with respect to indigenous people, articles 231 and 232 of the 1988 Brazilian Constitution state that the use of natural resources in indigenous lands can only be carried out with authorisation of the National Congress and requires a hearing with affected communities, and their participation in project results.

Recently the Brazilian government started requiring hydroelectric generators to implement four basic measures to compensate for environmental liabilities: invest in efficiency and move away from huge scale projects; decrease transmission and distribution losses; invest in environmental improvements such as forestation and soil conservation; and use water in a sustainable manner.

Should consideration of these negative impacts be incorporated in the model? And if so, to what extent is that possible? Some of them are already accounted for in the model. For instance, the irrigation scenario assumes that irrigated land in the absence of the dam would be substituted for by rainfed land, rather than retired. This is equivalent to assuming a one-to-one ratio of newly irrigated perimeters to the sum of flooded or unusable downstream land plus previously rainfed areas that are later irrigated. On the other hand, lost livestock production for displaced populations and *'vazanteiros'* could have been accounted for in the scenarios if accurate data were available. The same is true for lost fish harvest and their values, which could have been incorporated as an increase in crop production—the sector under which fishing activities are generally subsumed—in the without project scenario, thus reducing the impact of the dam in the with project scenario.

On the other hand, the model takes a Northeast-wide regional perspective, and is therefore, unable to account for intra-regional imbalances, such as the concentration of benefits from hydropower in coastal areas. Such impacts need to be accounted for in a parallel analysis of distributional impacts of the studied dams. Beyond such spatial considerations, a parallel distributional analysis needs to be performed anyway, given the inability of I/O-based models to trace the impacts on different income groups. Thus, nothing can be said about the social desirability of the value-added and employment creation resulting from the model simulations, as nothing is known regarding which income groups will benefit most from it. While some of the people that were initially negatively affected by the construction of the dams might later have benefited from its direct and indirect benefits, this cannot be simply assumed and cannot be explicitly derived from the model. Note, however, that in a second phase of the analysis, it should be possible to develop a SAM for the region whose labour factor use and household categories are disaggregated in ways that allow a SAM-based model to incorporate at least some of these issues.

Similar considerations can be made for environmental impacts. Given available data and the model's disaggregation, it was not possible to incorporate their consideration in the analysis. This is not an absolute impossibility however. As was stated above, even without refining the model, accurate data on reduced fish harvests and their values could have been incorporated in the scenarios. This, however, does not exhaust the total value of lost fish population to the ecosystems, which would have to be separately assessed. Intra-regional disaggregation of some accounts or a full spatial specification of the model, on the other hand, would allow the incorporation of impacts that are location-dependent, such as those from changes in flow regimes, sediment transport or increased pollution from the growth of urban centres in development *pólos*.

The possibility to refine/extend the model does not exclude the need for parallel social and environmental assessments of dam projects, particularly for the consideration of impacts that cannot be translated in economic terms. The issue here is simply to suggest that social and environmental assessments and the economic evaluation of impacts should be more carefully linked, so that the aspects of the former then can be incorporated in the economic analysis and are actually incorporated. This would enable a comparison of the full set of (additional) negative impacts on affected populations with the full set of positive impacts from the project, including some of the indirect impacts that may in fact benefit the initially negatively affected populations.

Conclusions

This case study had two objectives. One was to provide a characterisation and estimation of the direct and indirect impacts of the dams on the Sub-Médio São Francisco, given the claims made by many that the dams failed to deliver what they promised. The second was to emphasise some of the contextual factors that might have affected their capacity to produce multiplier effects. The capacity of a dam to generate real indirect and induced benefits depends largely on how its outputs build on the strengths and reduce the weaknesses of the regional context. It is only by recognising driving and constraining factors, and by designing the project and ancillary programmes accordingly, that a dam's potential multiplier effects might be realised.

The simulation results of a supply-constrained semi input-output model for Brazil's Northeast macro-region show that the large dams located in the Sub-Médio São Francisco have generated significant indirect and induced effects in the region. The results show that the dams may have accounted for 1.2 to 6 per cent of Northeastern GDP, depending on which 'without project' scenario is deemed more realistic. Value-added Type II multiplier values are close to 2.0 in most scenarios. For every unit of value generated by the sectors directly affected by the dams, another unit could be generated in the region. These secondary impacts are largely due to income-expenditure induced effects, however, as the effects of pure inter-industry linkages are much lower, with Type I multipliers ranging from 1.1 to 1.16. This result is in line with what other authors have reported regarding the dependence of the productive structure of the Northeastern region on demands from the rest of the country. With respect to employment, the range of multipliers is larger than for value-added, showing that conclusions regarding employment creation by dams are sensitive to the choice of scenario. Employment multipliers from agriculture are lower than one would expect based on other authors' findings regarding induced employment from irrigation. Again, this may descend from the model's incapacity to fully capture the impacts of dams on the structural changes irrigation induces in crop production and its links with the agribusiness productive chain. Thus, these results are bound to underestimate the full set of indirect impacts produced by irrigation, not only with respect to employment creation, but with regard to inter-industry linkages and Type I value-added multipliers as well.

Beyond the magnitude of these impacts, what matters is the information they provide regarding the structure of benefits that can be attributed to the dams. The dams were built primarily to feed the hunger for electricity of the Northeast's industrial complex. Due largely to distorted macroeconomic and industrial policies from the late 1960s to the mid-1980's, its excessive capital intensity and lack of significant inter-sector linkages may have impaired the dams' capacity to generate the larger impacts that one would have expected.

The model shows that while relatively large hydropower output shocks produce larger overall value-added than simulated reductions in irrigated area, these differences do not appear as large when one considers how small irrigated areas and their changes are compared to the region's planted area. In addition, the irrigation of these small areas produced the same. This is true even when the supply constraints on capital-intensive and energy-intensive sectors are removed. The results with respect to employment are even more striking, as the switch to irrigation of small areas produces the same percentage change in employment as large reductions in hydropower output, under the supply-constrained assumption. In addition, the large impact from hydropower descend primarily from income-expenditure effects on service demand, which may be overestimated due to the fixed price nature of the model. In the absence of the dams, alternative, more expensive sources of electricity would have been brought on board, which may have favoured a restructuring of the industrial complex in favour of less energy-intensive sectors. These might well have produced larger inter-sectoral linkages, in the end benefiting the region.

These results suggest that larger overall impacts might have been achieved if more attention had been given to multipurpose use of their water, while irrigation was given less prominence in the initial design of the project and in the design of resettlement and compensation programmes. Only in the late 1980's did macroeconomic and agricultural policies evolve in a way that recognised the role of agriculture and agribusiness as a motor of growth for the region.

Social and environmental impacts generated by the construction of the cascade of the reservoir, have been the object of serious scrutiny over the past decade. GDP growth associated with the dams' construction produced serious negative social and environmental externalities and, at least initially, negative distributional impacts. Hydropower development benefited primarily the already favoured coastal metropolitan regions, where energy-intensive industries tended to concentrate. Even the recent successes attributable to the modernisation of agriculture, its link with agribusiness and the spillover effects from the interactions of small and large farmers followed years of peasant displacement and increasing land concentration. Resettled peasants were often given plots with poor soils or in areas that required agricultural techniques they were unfamiliar with. Current positive impacts started materialising only after addressing the severe deficiencies of early interventions—such as the lack of technical support and ancillary infrastructure or other support programmes—which took more than a decade. For peasants who became landless, employment in rural areas was slow in emerging, while industries were generally in distant areas and many of them either required little labour—due to the bias in favour of capital-intensive industrialisation of ISI policies—or skilled labour for which they had no training. In addition, the irrigated area boom has not been reflected in a marked improvement in other areas of the Northeast, which may result in a worsening of intra-regional imbalances. The lesson learnt from experience is that appropriate parallel interventions need to be undertaken to minimise social externalities of dams, and that these may at the same time improve their economic outcomes.

While the region's economic performance is improving, its sustained development will depend on how effectively the Northeast addresses its environmental, socioeconomic, scientific, technological, political and institutional sources of vulnerability.

The isolated and at times, conflicting interventions that characterised public investment and other interventions in the Northeast, while contributing to structural change, did little to ensure sustainable social development in the region. Despite recent advancements in a number of indicators, income gaps kept widening

If the question of whether to invest in dams is posed as one of best use of limited funds, it is not clear whether it was the best decision to concentrate such a large amount of resources to benefit (i) an ill-conceived industrial complex, and (ii) a very small irrigated area which is only now showing its full growth potential. At the same time, the limited indirect impacts of the dams seem to have descended mainly from macroeconomic and other existing distortions. Yet, the lesson learnt from this is that dam projects should be designed with a clear understanding of the impacts these contextual elements have on the sectors they benefit. In addition, dams are large enough investments that their approval should warrant a re-evaluation of existing policies that may harm their outcomes. Alternatively, whenever possible they should be accompanied by parallel programmes that minimise the impacts of distortions, minimise social impacts, and increase the capacity of all potential beneficiaries—including otherwise negatively affected populations—to profit from the dam's development potential.

References

Accarini, J.H (2002). *"Pólos de Desenvolvimento Integrado Pólos de Desenvolvimento Integrado do Nordeste Brasileiro e Capital Social", Revista Econômica do Nordeste* 33(1). Fortaleza, CE.

ANA/GEF/UNEP/OAS (2002). *Integrated Management of Landbased Activities in the São Francisco Basin Project: Subsidies for the Formulation of Operational Policies for the Great Reservoirs in the São Francisco River Basin.* Final Report. Brasília, DF: ANA-Agência Nacional de Águas.

Aylward, B., J. Berkhoff, C. Green, P. Gutman, A. Lagman, M. Manion, A. Markandya, B. McKenney, K. Naudascher-Jankowski, B. Oud, A. Penman, S. Porter, C. Rajapakse, D. Southgate and R. Unsworth (2000). *Financial, Economic and Distributional Analysis*, Thematic Review III.1. Prepared as an input to the World Commission on Dams, Cape Town. Available at *www.dams.org*

Baer, W., M.A.R. Fonseca and J.J.M. Guilhoto (1987). "Structural Changes in Brazil's Industrial Economy, 1960–1980", *World Development* 14: 275–286.

Barros, F., (ed.) (2001). *Banco Mundial, Participação, Transparência e Responsabilização: A experiência brasileira com o Painel de Inspeção*, Rede Brasil sobre Instituição Financeira Multilateral.

Bell, C. and S. Devarajan (1980). "Semi-Input-Output and shadow prices: A Critical Note", *Oxford Bulletin of Economics and Economics and Statistics* 42(3): 251-56. August.

Bhatia, R., M. Scatasta, and R. Cestti (2003). "Study on the Multiplier Effects of Dams: Methodology Issues and Preliminary Results." *Presented at the Third World Water Forum*, held at Kyoto, Japan, March 16-23, 2003.

Bonelli, R. (2001). *"Impactos Econômicos e Sociais de Longo Prazo da Expansão Agropecuária no Brasil: Revolução Invisivel", Texto para Discussão No.* 838. Rio de Janeiro, RJ: IPEA.

Braga, B., O. Rocha, and J. Tundisi (1998). "Dams and the Environment: The Brazilian Experience", *Water Resources Development* 14(2): 127-140.

Cattaneo, A. (2002). *Balancing Agricultural Development and Deforestation in the Brazilian Amazon*. Research Report No. 129, Washington D.C.: IFPRI.

CODEVASF (2000). *Cadastro do Vale do São Francisco— 2000.* Available at *www.codevasf.gov.br*

———. (1999). *Inventário dos projetos de irrigação.* III ed. Brasília, DF.

———. (1996a). *Projetos de Irrigação no Vale do São Francisco,*.Brasília, DF.

———. (1996b). *Projeto Semi-Árido: Proposta de Desenvolvimento Sustentável da Bacia do Rio São Francisco e do Semi-árido Nordestino.* Brasília, DF.

———. (1989). *Plano Diretor para o Desenvolvimento do Vale do São Francisco — PLANVASF: Programa para o Desenvolvimento da Irrigação (1989-2000).* Brasília, DF.

Considera, C.M. and M.H. Medina (1998). "PIB por Unidade da Federacao: Valores Corrented e Constantes—1985/1986". Texto para Discussao 610. Rio de Janerio: IPEA.

Correia, R.C., J.L.P. Araújo, and É. de Barros Cavalcanti (2001). *"A Fruticultura como Vetor de Desenvolvimento: O caso dos municípios de Petrolina (PE) e Juazeiro (BA)", XXIX Congresso Brasileiro de Economia ae Sociologia Rural, Competitividade e Globalização Impactos e Regionais e Locais.* SOBER/ESALQ/EMBRAPA/UFPE/URFPE, Recife, PE.

ELETROBRÁS (1994). *PLAN 2015/National Electric Energy Plan 1993/2015*. Brasília, DF.

Ferradás, C.A. (2001). "Report of Social Impacts of Dams: Distributional and EquityIssues—Latin American Region", Contributing Paper Prepared for *Thematic Review I.1: Social Impacts of Large Dams Equity and Distributional Issues*. Cape Town: World Commission on Dams. Available at *www.dams.org*.

———. (1998). *Power in the Southern cone Borderlands. An Anthropology of Development Practice*. Westport, CT: Bergin & Garvey.

Ferreira, F. H.G., P. Lanjouw and M. NERI (2001). *A Robust Poverty Profile for Brazil Using Multiple Data Sources*, Paper presented in LACEA 2000 and ANPEC 2001 meetings.

FIBGE (1999). *Resolução nº 5, de 25 de agosto de 1999. Estabelece estimativa da população em 01/07/99*. Diário Oficial [da República Federativa do Brasil], nº 166, pp.81-108, 30 de agosto. Brasília, DF.

Filho, Pego Bolivar et al. (2001). *"Impactos Fiscais da Crise de Energia Electrica", Texto Para Discussão 816*. August. Rio de Janerio: IPEA.

———. (1991). *Censo Demográfico 1991: Resultado Final*. Rio de Janeiro, RJ.

França, F.M.C. and T.M.L. Maia (2002). *Pólos de desenvolvimento integrado: avaliação qualitativa de desempenho*. Fortaleza: Banco do Nordeste/IICA.

Gasques, J.G. (2001). *"Gastos Públicos na Agricultura", Texto para Discussão No.* 782. Rio de Janeiro, RJ: IPEA.

Gasques, J.G. and C.M. Villa Verde (1995). *Prioridades e Orientação dos Gastos Públicos em Agricultura no Brasil*, Texto para Discussão no.365, IPEA, Rio de Janeiro, RJ.

Guilhoto, J.J.M. (1998). *Análise Inter e Intra-Regional das Estruturas Produtivas das Economias do Nordeste e do Resto do Brasil: 1985 e 1995 Comparados*. Departamento de Economia e Sociologia Rural, ESALQ- USP. (mimeo).

Guilhoto, J.J.M., G.J.D. Hewings, M. Sonis and J. Guo (2001). "Economic Structural Change Over Time. Brazil and the United States Compared", *Journal of Policy Modeling* 23: 703–711

Guilhoto, J.J.M, M. Sonis and G.J.D. Hewings (2000). *Productive Relations in the Northeast and the Rest of Brazil Regions in 1995: Decomposition & Synergy in Input-Output Systems*, REAL 00-T-3, Regional Economics Applications Labouratory (REAL), the University of Illinois and the Federal Reserve Bank of Chicago.

Gutnam, P. (2001). "Some Evidences on Overall Distribution and Equity Impacts", Contributing Paper Prepared for *Thematic Review I.1: Social Impacts of Large Dams Equity and Distributional Issues*. Cape Town: World Commission on Dams. available at *www.dams.org*

Haddad, E.A., and G.J.D. Hewings (2000). "Linkages and Interdependence in the Brazilian Economy: An Evaluation of the Inter-Regional Input-output System, 1985", *Revista Econômica do Nordeste* 31(3): 330-367. Fortaleza, CE.

Haggblade, S., J. Hammer, and P. Hazell (1991). "Modeling Agricultural Growth Multipliers", *American Journal of Agricultural Economics* 73(2): 361-374.

Haguenauer, L., L. Dias Bahia, P. Furtado de Castro and M.B. Ribeiro (2001). *"Evolução das Cadeias Produtivas Brasileiras na Década de 90", Texto para Discussão no.* 786. Rio de Janeiro, RJ: IPEA.

Hall, Anthony L. (1978). Drought and Irrigation in North East Brazil. London: Cambridge University Press.

Harris, B., P. Hazell, and R. Slade (1987). "Regional Growth Linkages from Agriculture", *Journal of Development Studies* 23(2): 274-94.

Hazell, Peter B.R. and C. Ramasamy (1991). *The Green Revolution Reconsidered: The Impact of High-Yielding- Rice Varieties in South India.* Baltimore, MD: The Johns Hopkins University Press.

Hewings, G.J.D. (1985). *Regional Input-Output Analysis*. Beverly Hills: SAGE Publications

Leontief, W. (1970). "The Dynamic Inverse", in Carter and Brody (eds.), *Contribution of Input-Output Analysis.* Amsterdam, North Holland

Leontief, W. *et al.* (1953). *Studies in the Structure of the American Economy*. New York: Oxford University Press,.

Lima, J.P.R. and E.A de A. Miranda (2001). Fruticultura Irrigada no Vale do São Francisco: Incorporação Tecnológica, Competitividade e Sustentabilidade, *Revista Econômica do Nordeste*, v. 32: 611-632, Fortaleza, CE.

Maia Gomes, G. and J. R. Vergolino (1995). *"A Macroeconomia do Desenvolvimento Nordenstino: 1960/1994", Texto Para Discussão No.* 372. Rio de Janeiro, RJ: IPEA.

MMA/SRH (1999). *Anais do Ciclo de Palestras da Secretaria*. Brasília, DF.

Muller, A. C. (1995). *Hidrelétricas, meio ambiente e desenvolvimento.* Makron Books, São Paulo, SP.

Nishizawa, T. and J. I. Uitto (1995). *The Fragile Tropics of Latin America: Sustainable Management of Changing Environments.* Tokyo, New York, Paris: United Nations University Press.

Nordeste, Banco do (1998). *Documento Básico do Projeto Pólos de Desenvolvimento Integrado do Nordeste. Versão Preliminar.* Fortaleza, CE.

Operations Evaluation Department (1991). *Dynamics of Rural Development in Northeast Brazil:* New Lessons from Old Projects. Report, Washington D.C.: The World Bank.

————. (1998). *Recent Experiences with Involuntary Resettlement: Brazil-Itaparica*, No. 17544. Washington D.C.: The World Bank.

Pêgo Filho, B., J. Aroudo Mota, J.C. Jacob de Carvalho, M. Mota Saboya Pinheiro (2001). *"Impactos Fiscais da Crise de Energia Elétrica: 2001 e 2002", Texto para Discussão No*. 816. Rio de Janeiro, RJ: IPEA.

Prochnik, V. and L. Haguenauer (2001). *Cadeias Produtivas e Oportunidades de Investimento no Nordeste Brasileiro*, Presented at the XIV Congresso Brasileiro de Economistas, Recife, PE, September.

Rodrigues, L. (2001). *"Potencial da Agricultura Irrigada como Indutora do Desenvolvimento Regional: o caso do projeto Jaíba no Norte de Minas Gerais", Revista Econômica do Nordeste* 32(2): 206-232, Apr.-June. Fortaleza, CE.

Sampaio, Y. (1994). "Experiências de desenvolvimento rural e seus ensinamentos para o Nordeste do Brasil", *Revista Econômica Nordeste* 25(4): 529-544. Fortaleza, CE.

Soares, F. A. & Pereira, R.A.C. (2000). *"Modelo Multiregional: O Nordeste Brasileiro e Suas Relações Inter-Regionais", Estudos Econômicos No.*18. Fortaleza, CE: CENER, CAEN-UFC.

Stone, R. (1985). "The Disaggregation of the Household Sector in the National Accounts", in G. Pyatt and J.I. Round (eds.), *Social Accounting Matrices: A Basis for Planning.* Washington D.C. :The World Bank.

SUDENE (1999). *Portaria nº 1.182, de 14 de setembro de 1999. Anexo III—Região Semi-Árida*. Recife, PE.

———. (1997). *Região Nordeste do Brasil em números*. Recife, PE.

Ventura Filho, A. (2002). *The Hydroelectric Power Option in Brazil: Environmental, Technological and Economic Aspects*. World Energy Council. *www.worldenergy.org*

Vergolino, T.B. and J.R. Vergolino (1997). *"Reforma do Estado: Efeitos sobre a Renda e o Emprego na região do Sub-Médio São Francisco (Petrolina/Juazeiro)", Revista Econômica do Nordeste*, No. especial 28: 447-459, "Economia Agricola, Recursos Naturais e Meio Ambiente". Fortaleza, CE.

Watts, D. and R. Ariztia (2002). "The Electricity Crisis of California, Brazil and Chile: Lessons to the Chilean Market", Presented at LESCOPE' 2002, *The 2002 Large Engineering Systems Conference on Power Engineering*. Halifax, Nova Scotia, Canada, June.

WCD (2000). *Dams and Development: A New Framework for Decision-Making*. The Report of the World Commission on Dams. Issued Nov. 16. *www.damsreport.org*

World Bank (1996). *The World Bank's Experience with Large Dams—A Preliminary Review of Impacts*, Operations Evaluation Department Report No. 15815. Washington D.C..

———. (1998). *Recent Experience with Involuntary Resettlement: Brazil—Itaparica*, Operations Evaluation Department, Report No. 17544. Washington D.C..

———. (1994). *Project Completion Report, NE Irrigation Engineering and Technical Assistance Project*. p.1

World Energy Council (2003). *The Challenge of Rural Energy Poverty in Developing Countries*. U.K.- World Energy Council.*www.worldenergy.org*

Annex A-9.1

The 1992 Input-Output Table: 30x30 Activities Version (million 1992 Cruzeiros)

	AAGR (1)	APEC (2)	AEXMIN (3)	AMAMIN (4)	ASID (5)	AMECH (6)	AELETR (7)	AELETRON (8)	ACARBUS (9)	ATRAMAT (10)	AFURN (11)	APAPULP (12)	AGRAPH (13)	ARUBB (14)	ACHEM (15)	APERF (16)	APLAST (17)
AAGR	2254324	700161	928	14169	71491	46	102		61	15	65119	23521	12767	3531	1291539	4146	89
APEC	639378	501979	99	1224	6261	10	34		13	3	5940	2141	1149	306	116592	1004	17
AEXMIN	95279	40789	28537	154893	119558	253	3406		323	92	145	3793	1932	144	3220886	2301	377
AMAMIN	5461	2064	33155	584316	40909	1121	24292		1042	309	3047	1601	1173	57	177209	8454	1461
ASID	17410	6818	134041	75897	4510497	41919	344035		52942	13601	15037	20854	9569	1015	185155	5619	7530
AMECH	5200	2035	65861	5960	36503	2016	11828		1062	585	561	2297	1159	146	45509	454	1004
AELETR	951	174	2185	1084	27393	4692	217803		1867	324	239	584	212	38	10046	90	413
AELETRON	7	2	70	108	280	31	306		12	3	6	19	9	1	279	4	7
ACARBUS	491	46	440	132	2887	365	442		4932	31	105	52	24	24	875	14	26
ATRAMAT	93	18	428	86	1582	488	179		23	4835	16	34	17	20	661	8	14
AFURN	6195	2464	1201	1305	10934	190	2544		445	133	30112	3308	1611	26	4553	81	429
APAPULP	13571	5824	2863	62466	79584	1218	27124		1473	10	5417	253370	189726	346	161354	28480	14487
AGRAPH	490	217	92	1412	4693	68	728		66	1	221	6177	14052	8	3746	1137	981
ARUBB	105	29	665	1501	6634	566	2541		812	100	899	1081	551	8255	6719	332	671
ACHEM	1413374	426215	54178	130216	722946	2841	77191		4746	1287	33610	111525	54281	21044	6719382	93409	331437
APERF	20588	3508	251	339	1978	11	136		10	3	79	343	170	11	6371	3284	93
APLAST	9424	23065	2437	3897	14006	766	21935		1236	310	11307	3403	1694	130	25737	6522	25667
ATEXT	54972	22846	4243	8725	20015	553	2520		1155	324	9380	6092	3181	5247	48932	1040	10008
AVEST	556	164	3080	1026	10390	95	703		97	26	299	1365	500	165	11809	295	302
ALEATH	7362	2177	377	338	2220	108	453		61	12	1330	476	224	291	4397	134	355
AFOODPR	655461	288178	4500	4330	37986	363	1503		367	98	1656	9274	3253	86	331454	31543	1089

contd...

contd...

	AAGR (18)	APEC (19)	AEXMIN (20)	AMAMIN (21)	ASID (22)	AMECH (23)	AELETR (24)	AELETRON (25)	ACARBUS (26)	ATRAMAT (27)	AFURN (28)	APAPULP (29)	AGRAPH (30)	ARUBB (31)	ACHEM (32)	APERF (33)	APLAST (34)
ADRINK	1042	291	174	185	1235	7	99		8	2	45	126	59	3	24943	199	69
AOTH	4972	1448	3732	3258	140390	211	2788		219	39	775	20609	5477	375	38495	408	1965
ASIUP	195267	29048	295382	167856	1080950	6520	47559		6638	1829	23119	75895	37561	2565	1718335	14650	25108
ACONSTR	968	302	27797	5592	19796	495	2800		399	107	1132	3501	1784	103	51460	1408	1082
ACOMM	1116778	38788	139202	158369	484701	8481	91819		7770	2065	38714	88816	43039	3374	1445491	52171	24135
ATRANS	492109	335	62038	69237	200861	3835	43218		3658	982	18175	37529	19140	1547	542076	24992	11233
ACOMMUN	5535	2154	14694	28005	68089	16309	16015		7689	1705	17513	30583	17396	4371	64983	28099	8010
AFININS	16380	7020	7603	4379	16982	1033	2715		1623	360	1102	3047	1750	512	18584	275	632
AOTHSVC	341465	137621	509946	80662	195462	32816	43387		7250	3676	44346	56333	57488	4240	298320	59553	16999
MROB	3000709	961088	79704	418683	1912311		335947			15327	58857		82458		2950152	65895	12170
MROW	345391	39319	70832	106859	697712	160544	22562	5462	56147	10301	14441	48261	27916	47318	1724561	62995	39298
CTAX	611745	211951	39683	66554	174751	11317	21695	4025	9536	1758	7771	17663	10227	9951	347940	31001	10894
REMUN	3777155	1542782	571336	513521	688872	170785	194961	1654	35451	65601	115583	99077	266093	25448	1616046	94802	130022
PROF	17997240	9273506	2567477	1271990	2370278	186383	571635	2017	127431	99574	217658	243267	235935	30574	6856760	218720	211988
ATAX	804	329	51467	44006	146461	7862	23392	144	2937	1895	8388	13212	12398	1846	314686	9190	10196
ASUB	-105846	-43233			-350	-329	-537		-82566	-63921		-23798	-22332				
OUTP	33002410142315204780697399257913927250663989						2159864	13302	256934	163404	752145	1165433	1093644	173119	30386030852709		900260

contd...

contd...

	ATEXT (35)	AVEST (36)	ALEATH (37)	AFOODPR (38)	ADRINK (39)	AOTH (40)	ASIUP (41)	ACONSTR (42)	ACOMM (43)	ATRANS (44)	ACOMMUN (45)	AFININS (46)	AOTHSVC (47)
AAGR	390132	743	9496	5635473	230501	2508	14082	7350	378	8406	1	1	774885
APEC	106126	198	2100	1347953	34455	263	1221	2885	156	1495	0		108995
AEXMIN	736	260	1120	21646	15233	33591	2868	179422	647	123	32	88	1371
AMAMIN	2239	280	652	82807	19789	6795	3566	2534992	1067	522	2520	121	99240
ASID	36902	8102	4844	145572	18925	52693	42793	1139246	22107	20340	9798	8873	107344
AMECH	16291	607	756	27708	1685	1105	64391	179113	10365	4360	2468	254	16331
AELETR	2145	289	252	3005	307	7019	633887	1014524	12906	3528	41885	644	100388
AELETRON	185	6	8	220	16	42	97	148	14	52	1305	2	1673
ACARBUS	383	48	83	399	46	52	1713	3288	334	69150	1083	29	99131
ATRAMAT	232	19	63	321	32	27	1103	1389	238	31212	33	11	2338
AFURN	2644	673	2214	7576	1493	3707	1698	249248	12274	3562	93	469	18886
APAPULP	92183	19099	35368	357055	39479	28189	34440	24478	378360	15622	26562	94778	588116
AGRAPH	2724	1122	861	11116	8048	818	20572	2312	109140	10683	554	6351	211456
ARUBB	13127	1394	21163	2624	185	1788	2312	13727	97	32699	183	15	52589
ACHEM	1366069	23593	90767	244421	60637	74449	364036	434497	1021390	92590	15108	2496	601277
APERF	2143	56	175	4256	2338	127	1845	1571	617	167	30	108	78259
APLAST	45642	7382	27806	59688	10258	12170	4526	226946	76946	96531	5566	368	77693
ATEXT	4942440	1133346	28652	274446	5065	19573	33155	14052	48232	86287	679	2545	431753
AVEST	11950	3867	2486	8446	704	655	7220	6747	3059	2134	1358	1052	18467
ALEATH	6448	17293	160948	5529	1255	2477	3990	2824	1583	2144	1817	176	16502
AFOODPR	18847	3170	68191	2739117	487671	3255	40433	11986	29072	45405	1114	6400	1015256
ADRINK	561	70	144	67366	51786	65	1605	975	16195	660	258	70	908981
AOTH	11416	8352	2956	20195	1583	7949	87947	18078	11178	4450	1860	25057	99987
ASIUP	487764	45554	43471	584103	71750	23643	3355080	97685	727421	90730	28220	46610	1503767

contd...

contd...

	ATEXT (48)	AVEST (49)	ALEATH (50)	AFOODPR (51)	ADRINK (52)	AOTH (53)	ASIUP (54)	ACONSTR (55)	ACOMM (56)	ATRANS (57)	ACOMMUN (58)	AFININS (59)	AOTHSVC (60)
ACONSTR	14816	1992	1766	39695	3066	1460	177713	1335060	79810	68799	27107	19	1255455
ACOMM	548934	90922	50483	1136881	178940	51360	310146	1446445	1787803	890815	71548	124184	2656634
ATRANS	263794	42446	24035	546243	86752	22467	95794	675375	689025	313328	30477	45099	1032391
ACOMMUN	30249	13256	11889	97726	11920	8243	63940	38994	280244	101783	6653	362808	459009
AFININS	7138	778	875	18993	1776	1423	66125	8924	42791	27193	7348	145272	84635
AOTHSVC	179842	70478	35686	463392	68823	126501	1284778	654780	2911690	632997	272029	1613277	5788635
MROB	976017	94596	117947	2185125	441161		1733570	4050389	5573815	2926482	201744	882010	4887001
MROW	164870	18045	16027	884921	66422	20868	766325	141426	328899	2784796	35966	84025	512761
CTAX	127594	16503	13836	262957	35394	8959	715495	857421	986988	869427	59941	246650	2263146
REMUN	1090619	353182	162993	2615485	385306	152498	5267079	4237531	10339290	3120248	1228554	9038017	43994200
PROF	2976431	702736	255939	5384889	556813	348101	4828294	15044560	13803120	3543674	1559561	18940630	32413190
ATAX	148371	30321	13125	273116	32787	11345	254686	382456	496423	193847	46548	407567	1002094
ASUB	-359121		-69498	-91835			-3358	-14927	-301933	-367627	-15790		
OUTP	13728880	2710776	1139680	25468630	2932399	1036185	20285170	35025910	39501740	15728620	3674215	32086080	103283800

contd...

contd...

	DUMMY (61)	*HHCONS (62)*	*INV (63)*	*DST (64)*	*EROB (65)*	*EROW (66)*	*TOTD (67)*
AAGR		7475857	754615	-907052	13073430	1089587	33002410
APEC		1837494	65415	2262349	7034573	149692	14231520
AEXMIN		7318	318	-502730	1148604	197339	4780697
AMAMIN		265515	7364	-113717	156649	36507	3992579
ASID		474010	129415	-988662	4148179	3104827	13927250
AMECH		67531	1400811	-1638610	316564	10078	663989
AELETR		488857	160068	-907811	285760	44116	2159864
AELETRON	1280	7260	-528	348	31	13302	
ACARBUS	29093	28404	-4093	11919	4986	256934	
ATRAMAT	7834	79361	-10	24320	6379	163404	
AFURN		266705	77589	5208	22314	10261	752145
APAPULP		273644	428	-2574590	857142	27768	1165433
AGRAPH		264302	5061	385386	18328	723	1093644
ARUBB		14405	493	-20737	5438	152	173119
ACHEM		3286628	-10760	-119273	9824971	2815455	30386030
APERF		709958	277	-25524	6207	32921	852709
APLAST		106178	6659	-103805	32255	55914	900260
ATEXT		1556954	4474	19509	4237905	690583	13728880
AVEST		2587986	1316	-33321	22320	33456	2710776
ALEATH		700807	194	-53453	12877	235952	1139680
AFOODPR		10936520	9201	-84551	5738071	3028338	25468630
ADRINK		1749762	424	21710	59126	24151	2932399
AOTH		357938	17761	21550	108803	3964	1036185
ASIUP		5295070	129451		4025770	801	20285170

contd...

contd...

	DUMMY (68)	HHCONS (69)	INV (70)	DST (71)	EROB (72)	EROW (73)	TOTD (74)
ACONSTR		3535	31838070	-49	58597	273	35025910
ACOMM		20476890	997632	132897	3530819	1274694	39501740
ATRANS		7775716	489171	-103683	1584385	584833	15728620
ACOMMUN	1705038	597		139533	11182	3674215	
AFININS	29430030	2131101	447	0	22949	4287	32086080
AOTHSVC		84276820	654368	308399	1639344	412433	103283800
MROB	80380430	48397960	6165931	6565792		2361118	9893301
MROW		1834199	849400	402606		9458	3479919
CTAX	1295261	14470360	2340138	-102757		648223	1588462
REMUN							9909189
PROF							42482433
ATAX							649213
ASUB							-342912
OUTP	111105700	219833300	46211350	1840450	58147500	16910480	

Source: Author's calculations based on Considera *et al.* (1997).

Where:

AAGR	Crop Cultivation
APEC	Livestock Production
AEXMIN	Mineral Extraction
AMAMIN	Non-Metallic Mineral Manufacturing
ASID	Steel Production
AMECH	Machinery
AELETR	Electric Products
AELETRON	Electronic Products
ACARBUS	Motor Vehicles
ATRAMAT	Other Transportation Equipment
AFURN	Wood, Furniture and Fixtures
APAPULP	Pulp and Paper
AGRAPH	Publishing and Printing
ARUBB	Rubber Production
ACHEM	Chemical Industry
APERF	Pharmaceuticals and Fragrances
APLAST	Plastic Production
ATEXT	Textile Production
AVEST	Wearing Apparel and Other Textile Production
ALEATH	Leather, Fur Skins and Miscellaneous Leather Production
AFOODPR	Food Processing
ADRINK	Drinks
AOTH	Miscellaneous Manufacturing
ASIUP	Provision of Public Utility Services—Electricity, Gas, WS&S
ACONSTR	Construction Activities
ACOMM	Commercial Activities
ATRANS	Transportation Services

ACOMMUN	Communication Services
AFININS	Finance and Insurance
AOTHSVC	Other Public Services
MROB	Imports from Rest of Brazil
MROW	Imports from Rest of World
CTAX	Commodity Taxes—include Custom Duties and Tariffs on Imported Goods
REMUN	Total Remuneration of Employees
PROF	Operating Surplus—includes Depreciation
ATAX	Indirect Activity Taxes
ASUB	(net of) Current Subsidies
OUTP	Total Domestic Production (= total gross inputs) by activity
DUMMY	Financial Dummy—Balancing Sector
HHCONS	Current Consumption (private + public)
INV	Gross Domestic Fixed Capital Formation
DST	Change in Stocks
EROB	Exports to Rest of Brazil
EROW	Exports to Rest of World
TOTD	Total Demand by Activity

Annex A-9.2

Full Set of 1992 Regional Input-Output Tables

The full I/O table comprises the following sets of information:

Table A.1: Regional Output

Presents the Value of Output at basic prices.[1] Each row represents output by final product of each productive process (activity), so that the row sum is the total value of output by activity. Each column represents the origin, by process, of each final product, so that each column sum represents output by final product (commodity).

Table A.2: Intermediate Consumption (by Activities) and Final Demand of Commodity Produced in the Region

This is generally thought of as the I/O table proper. Each row represents the destination of each commodity, valued at approximately the basic price. From this table, the model generates the product matrix by activity of intermediate consumption, the matrix of generated income, and the matrix of products for final demand. This Table also contains information on labour demand (employees) by activity.

Table A.3-4: Intermediate Consumption and Final Demand—Imported from the Rest of Brazil. Imported from the Rest of the World

These mirror Table 2, but only for inter-regional or international imports, valued at CIF. It corresponds to the disaggregation of the import from Brazil row in Table 2

Table A.5-7: Consumption Structure

Computed based on Tables 2–4, they result from dividing intermediate consumption by total value of production (net of imports and following rows in Table 2).

---

1. In billion 1985 cruzeiros and million of 1992 cruzeiros. The cruzeiro (Cr) is the currency that preceded Brazil's short-lived Cruzeiro Real, introduced in 1993, and the current Real, introduced in 1994. As 1990-1994 were characterised by hyperinflation, and given the recent depreciation of the real, it is misleading to use the official 2750*1000 Cr/R$ conversion rate suggested by various sources. The author preferred to maintain the values in 1992 Cr, given that what is of interest for the analysis are percentage changes and multiplier values, rather than absolute values.

Table A.8: Matrix of Regional Technical Coefficients

A square activity x activity matrix showing the intermediate consumption requirements by activity.

Table A.9: Leontitef Matrix

Computed based on Table 8, this Matrix shows direct and indirect impact coefficients, capturing the impacts on production of exogenous changes in final demand. Column totals represent backward linkages. Row totals represent forward linkages.

Table A.10: Impact on Intermediate Consumption Imported from Rest of Brazil

Results from multiplying Table 6 by the Leontief Matrix

Table A.11: Impact on Intermediate Consumption Imported from the Rest of the World

Results from multiplying Table 7 by the Leontief Matrix

Table A.12: Employment and Employment Creation Coefficients

Direct coefficients are obtained dividing employment by the value of production by activity. Total coefficients are obtained multiplying direct coefficients by the Leontief Matrix, and representing direct and indirect employment generation from an additional unit of final demand by each row activity.

Table A.13: Market Share Table ("Make Matrix")

It indicates the sectoral origin (by activity) of each final commodity

Table A.14: Distribution Margins Table (1985)

It has the same structure as Table 2 and identifies marketing margins and transportation costs associated with intermediate or final consumption.

Table A.15: Table of Direct and Indirect Taxes, and Subsidies (1985)

It identifies their incidence on intermediate and final demand.

10 Afterword

How Theory, Practice, Politics and Time Affects Views on the Indirect Economic Impact of Water Infrastructure.

John Briscoe

This book has described both how a new generation of analytic tools can be used for examining the direct and indirect benefits of major infrastructure in the developing world and presents results which consistently show that major water projects give rise to very large indirect benefits.

The purpose of this Afterword is to locate this theoretical and empirical work in a historical, political and economic context. It builds on the premise that analytic tools and analytic results do not develop in a vacuum, but that they are deeply affected by issues such as the stage of development, and the governance of the development process. It will be shown that the questions which are asked, and the answers which are accepted, depend on who is doing the asking, when they are asking and who is deciding what answers are acceptable. It will be shown that methods and conclusions which might be appropriate at one period in the historical evolution of a country are inappropriate at other periods in the development process. It will also be shown that whereas these political and contextual questions are of muted importance when there is a self-contained examination within a particular country (where moral hazards are limited) they are of transcendental importance in the development business (where it is usually rich countries defining what is acceptable for poor countries, and where post-development insights are imposed on pre-development contexts).

An instructive point to start our story is in 1955, when there was a heated discussion between the US Bureau of the Budget and the US Congress on the question of indirect benefits of major water infrastructure. The Congressional Record of these two discussions in 1955 (U.S. Congress, 1955) should be a required reading for all students of the history and political economy of the apparently-neutral tool of cost-benefit analysis!

The hearings were called to review a proposal by the Bureau of the Budget which, inter alia, suggested that indirect benefits should be excluded when evaluating major water infrastructure projects. At the first hearing the proposal was hotly contested by the Committee Chairman and members of the House Committee. (Some of the Chairman's utterances: "I want the gentleman (Director of the Budget Bureau) to have the milk of human kindness in his soul as he looks over the reasons this order was issued" and "Let the Chair suggest to the gentleman that a vicious policy, wherever it originates, should be stopped."). How, the Congress asked, could the government propose to exclude consideration of the indirect benefits, when the major reason why projects like the TVA and Grand Coulee had been built was precisely in order to generate such benefits? The irate Chairman (from California) asked the Director of the Bureau to consider what development in the Western United States would have been if the proposed methodology had been adopted in the developmental stage of US history. "Do you mean to say that the people in your agency set up a document which is to set the standards... And yet you cannot name one single project ever built by the Bureau (of Reclamation) or currently under consideration that would qualify for construction under the standards set in that document?"

In the second hearing, two weeks later, the Secretary of the Interior explained that "a basic purpose of the reclamation (i.e. major water) laws is to spur development in the West". There was, therefore, general agreement that indirect benefits were important. The problem, he further explained, was that "while there is general recognition that such (indirect benefits) are created by the projects, agreement has not been reached as to how such values are properly measured".

The result of the hearings was a substantial revision of the original proposal. As summarised in the standard World Bank book for the Evaluation of Agricultural Projects (Gittinger, 1982) "when market prices are used in economic analysis, as has been the custom in the US for water resource and other public works, it is necessary to estimate the secondary costs and benefits and then add them to the direct costs and benefits".

Heated though the discussion was, in many ways this debate was of waning practical importance in the United States, because by the 1960s the country had developed the infrastructure platform for its broad-based economic development. The US developed about 5000 cubic meters of water storage capacity per person, with 900 days of storage capacity on the Colorado River, and with about 75 per cent of hydropower potential developed. Most of this infrastructure was explicitly and deliberately designed to provide a platform for broad-based development. Where detailed subsequent assessments of indirect benefits were subsequently undertaken (for example of the Grand Coulee Dam (Ortolano and Cushin, 2003), these confirmed that the indirect effects were, in fact, very large, typically about the same size as the direct impacts of the energy and irrigation services provided.

But what of the developing world, where such questions are far from academic, since, almost by definition, very little of this infrastructure platform had been developed? Compared to the reservoir capacity of 5000 cubic meters per capita in the US, Ethiopia and Kenya can store only 30 cubic meters; compared to the 900 days of storage in the Colorado, Pakistan can store only 30 days of flow on the Indus; compared to the 75 per cent development of hydropower potential in the US, Africa has generated less than 5 per cent, and Asia and Latin America about 30 per cent. In short, what is the import of this debate when it takes place before the water infrastructure platform (Grey and Sadoff, 2007) is in place?

In the 1960s and 1970s there was intensive intellectual effort devoted to defining methodologies for assessing the economic impact of projects in the developing world. A major difference between the case of the United States (the focus of the Bureau of the Budget analysis) and the developing world was that markets were often severely distorted by governments in the developing world. To deal with these price distortions, economists developed methods for using shadow prices (which reflects the opportunity cost for inputs and intermediate outputs, and willingness to pay for final goods and services). For example, the definitive 1972 UNIDO treatise on cost-benefit analysis (UNIDO, 1972), described a four-stage process:

- Step One: Financial Analysis: calculation of commercial profitability at market prices;
- Step Two: Economic Analysis of Benefits: Calculation of net aggregate consumption benefit at shadow prices
- Step Three: Economic Analysis of Costs: Adjustment of the social value of investment
- Step Four: Other Goals: Addition of benefits deriving from the accomplishment of other national goals such as regional development and income distribution.

Subsequent elaborations (for example Gittinger, 1982) forged a link between the shadow price analysis (Steps 2 and 3) and Step 4, asserting that "Shadow prices that include carefully traced indirect changes in value added include the multiplier effects while minimizing the danger of double counting" and claimed that "most of the multiplier effect is accounted for if we shadow-price at opportunity cost" (Gittinger, 1982). These elaborations cautioned against "estimation of indirect benefits, which is a theoretically difficult process and one easily subject to abuse."

And there, to a substantial degree, is where things stopped, with general practice in development economics being use of shadow prices to estimate inputs and outputs, with indirect effects being ignored both (a) because it was asserted that these would be taken account of through careful shadow pricing and (b) because it was considered methodologically and computationally too difficult to estimate these indirect effects.

Chapter II of this book, co-authored by one of the world's leading practitioners of project cost-benefit analysis, reviews the way in which shadow pricing is actually applied in project analysis. This review casts serious doubts on the claim that using shadow prices actually results in an accurate assessment of costs and benefits. It shows, for the case of irrigated agriculture, that (a) the use of shadow prices for major outputs either overestimates or underestimates the value-added from the primary processing of agricultural commodities and (b) that conventional cost-benefit analysis usually confines its attention to primary outputs (such as paddy, wheat, cotton, sugarcane), seldom explicitly taking into account secondary forward linkages (such as conversion of wheat into bakery products or oils for hydrogenated oil or sugar into confectionery). The analysis in Chapter II concludes that, in practice, the use of shadow prices seldom captures the indirect or multiplier effects.

The use of shadow pricing methods has tended to fade in recent years as a result of very important changes in the economic landscape in most developing countries. Whereas heavy government intervention was standard in the 1960s, the level of government intervention, and the associated distortions, has fallen fast in most developing countries. Today, prices are generally, as in developed countries, reasonable reflections of the values of the inputs and outputs. This has meant that standard practice in project evaluation has thus converged on the long-standing practice in developed countries, where (as described in Gittinger, 1982) market prices are used directly in cost-benefit analysis. (To illustrate this change, the words "shadow" does not occur in the economic analyses done of the two major recent World Bank financed dam projects, Bujagali in Uganda and Nam Theun 2 in Laos).

Simultaneously computational capacity has improved dramatically, as has the availability of calibrated models of regional and national economies. As shown in the empirical sections of this book, it is now possible, in many settings and at modest cost, to estimate, ex post, the indirect costs and benefits of large projects and, in many cases, estimate which economic groups benefit from these indirect impacts.

A summary of the findings in this book (of indirect benefits of large dams in Egypt, Brazil and India), in other studies (of the Grand Coulee Dam in the United States (Ortolano and Cushin, 2003) and the Muda Irrigation Project in Malaysia (Bell and Devarajan, 1985)) show:

- Methodologies are now available which enable economists to estimate, *ex post*, the indirect effect of major water infrastructure projects and the incidence of these benefits on different groups.

- Indirect impacts are real and large—typically of the same order of magnitude as the direct effects—in the development stages in rich countries and in contemporary developing countries;
- The incidence of benefits is quite different when indirect effects are added to direct effects, with some striking cases (Bhakra in India) where it is the poor, who were not beneficiaries of the direct impacts, who were the major beneficiaries—through the labor market—of the indirect impacts;
- The sources of actual total (direct and indirect) benefits are very different ex post than was predicted ex ante. The most striking case is Aswan, where the logic of the dam was to develop Egyptian agriculture, but where by far the largest direct and indirect benefits have emanated from electricity generation. A deeper point is that a major attribute of multi-purpose dams is to produce major streams of benefits as economies evolve and as societal demands evolve.
- While the number of cases is small, the evidence in this study is that smaller dams (Bunga in India) have, as anticipated, smaller ripple effects and thus smaller indirect effects than large dams (Bhakra in India)
- While the literature is small, there is also evidence (Hazell and Ramasamy, 1991) that the multipliers for infrastructural investments are much larger than for the "social" investments which are currently given highest priority by development agencies.

The second part of this chapter examines how these striking results, confirming the transformative role of large water infrastructure, has been translated into contemporary development policy. One way of understanding this evolution is to tell the story from the perspective of the World Bank, which occupies center-stage in the development community, especially in terms of framing the intellectual debate about development.[1]

1 The author was an active participant in this debate in the World Bank, during his tenure as the Senior Water Advisor at the Bank during the period from 1995 to 2005. A description of the war over infrastructure policy in the Bank during that period has been described in detail in Chapter 13 of Sebastian Mallaby's The World's Banker, the definitive history of the Bank during this period (Sebastian Mallaby. 2004).

To understand how the development community has responded to these results, it is necessary to start with an understanding of the way in which major development institutions are governed, with the World Bank an instructive illustrative example. The World Bank lends to developing countries through two main "windows"—about 75 per cent of lending is done at market rates through the IBRD; about 25 per cent is done through heavily-subsidised rates through IDA to poor countries. One might expect that hard-lending-window thinking (IBRD) would dominate the intellectual landscape in the Bank, both because it is the longest-serving instrument but also because of its numerical dominance. The reality is just the opposite, for reasons described by the author of the official history of the World Bank. In a brilliant analysis, Devesh Kapur (2002) shows how, in recent decades the IDA tail has come to wag the IBRD dog. The roots of this paradox are a contrast in the politics of raising capital. Raising capital for the IBRD window is a quiet and largely invisible process of issuing bonds, and selling these to long-term institutional investors. By contrast, raising capital for IDA is a highly-political, permanent process in which the World Bank goes around to the capitals of the rich world with cap in hand. As with all charitable transactions, he who pays the piper calls the tune. In this case rich countries use the leverage of IDA to impose the view of their constituents who are interested in development on the World Bank. Central to this process is the rapid increase in activism by rich country NGOs who have either a charitable, paternalistic view of development or a profoundly anti-capitalist view of the development process. As described in Sebastian Mallaby's (2004: 8) seminal description of the recent history of the World Bank, rich-country NGOs have come to exert a huge influence on the development policies of their governments, and have often had a fundamental aim of re-shaping the World Bank to conform to their image of development. And what is this vision? In many ways it could be described as leaping directly from poverty to a welfare state, without the intermediate step of developing a productive economy. The Millennium Development Goals (MDGs) are the perfect articulation of this view—they prioritise services such as health, education, basic water and sanitation—but say nothing about the economic processes which have underpinned the development of such services in now-rich countries. The MDGs are quiet on the fundamentals of economic growth—on the investment climate, and on investments in critical institutional and physical infrastructure for agriculture, transport, energy and water resources.

The ironies in this process are astounding. The NGOs incessantly complain that "the developing world is not adequately represented in institutions like the World Bank" and incessantly complain about "the World Bank imposing conditionality on developing countries". Yet they specifically target instruments like IDA, where the developing world has no voice, to impose extreme conditions which the developing country members of the World Bank would never voluntarily accept. They focus exclusively and heavily on "sins of commission" ("was anyone adversely affected by this process?") and ignore the sins of omission ("how many people were harmed by not doing this project"?)

The recent debate over large dams at the World Bank provides an illustration of how this works in practice. In the early 2000s the staff of the World Bank advocated, for the first time in decades, that the Bank again finance politically-incorrect large dams. A major World Bank consultative process (World Bank, 2002) elicited views on the Bank's proposed re-engagement with "high-risk, high-reward" water infrastructure from different groups of "stakeholders". There was strong support for Bank engagement across the spectrum in developing countries – from governments, the private sector, academics and most NGOs. But there was strident opposition from rich country NGOs whose views were, strikingly, very close to those of representatives of rich country aid agencies. These divergences were not hidden but highlighted, in particular to the developing country members of the Board of the World Bank.

The debate at the Board was, for once, dominated by India, China, Brazil, and African countries, who strongly supported this policy change. The Executive Directors for the rich countries were uncharacteristically reticent in the face of the concerted and strongly-felt views of the developing country owners of the Bank. But the matter did not rest there, but descended to the subtle blackmailing which characterizes the formulation of development policy. As the World Bank Vice President in charge of this process returned to his office from the first of these exchanges at the Board, the representative of one of the major rich-country owners of the Bank, who had not said a word at the meeting, was on the phone. "If this is the position taken by the Bank, then you must realize that this will put our contribution to IDA in question." This is the way in which the IDA tail comes to wag the IBRD dog!

Sebastian Mallaby has described (Sebastian Mallaby, 2004) the toxic effects of this reality on the World Bank and the broader process of development. Jim Wolfensohn assumed the Presidency of the World Bank at the height of the red-green NGO onslaught on the Bank. One of the first issues he faced was a decision on Bank support for a medium-sized run-of the-river hydropower project (Arun) in Nepal. The project was assailed by the "there is no such thing as a good dam" NGOs. Wolfensohn had strong links with these NGOs and wanted to buy peace with these critics (and their rich country supporters on the Board of the Bank). Within the Bank there was a line of reasoning which said that the project was "too big" for the economy of Nepal (ignoring the fact that neighboring Bhutan, a fraction of the size of Nepal, had blossomed as a result of much larger-scale hydropower development). Using this fig leaf of economic analysis, Wolfensohn's first major symbolic act as President of the World Bank was to withdraw Bank support for the Arun project.

With Wolfensohn's support, the Bank subsequently joined with the IUCN to launch what was intended to be a broad-based multi-stakeholder effort to define new standards for the construction of large dams. What happened was that the World Commission on Dams (WCD) was hijacked by the anti-dam NGOs "who have no off switch" (Sebastian Mallaby, 2004). In a remarkable, gloating piece (McCully, 2005), the NGO leader of this capture described in detail, for the benefit of other such efforts in the future, how the radical NGOs had no intention of compromise, how they were able to ensure that developing country governments had no voice, how they were able to blackmail moderate NGOs and how they now intended to shove these standards down the throats of developing country governments who the NGOs had effectively excluded from the WCD process. A central analytic decision implicit in the WCD report was that only direct impacts of large dams should be counted as benefits. The final WCD report prescribed guidelines which would, if followed, mean that no large dam could ever be built again.

To every reaction there is, eventually, a reaction. When the Board of the World Bank considered the WCD report, there was vociferous opposition from many developing countries to the Bank adopting the impractical WCD guidelines. From the perspective of this chapter, it is relevant to note that an important element was the perception of how indirect benefits were dealt with in the report of the WCD. Just as there had been deep disquiet in the 1950s in the US Congress about OMB's proposal to exclude indirect benefits, so too there was now disquiet among developing countries about the application of the "indirect effects do not count" philosophy in developing countries. Developing countries noted, repeatedly, that major water infrastructure had been a platform for the development of most now-rich countries, and that the major emerging developing countries—India, China, Turkey, Brazil—had all invested heavily in major infrastructure. Developing countries "with choices" (such as India, China, Turkey and Brazil) were unequivocal that they would continue to build major water infrastructure. Developing countries "without choices", who depend on the goodwill of the World Bank and other development partners, insisted that it was immoral for rich countries who had such infrastructure deny them the right to develop.

The unwillingness of the developing countries to have their voice stifled (led by the "countries with choices") set the stage for a new World Bank policy which emphasised that institutional reforms were necessary but that, so too was major water infrastructure (World Bank, 2003). The increasingly self-confident developing country members of the Board of the World Bank, led by India and China, insisted—successfully—that major infrastructure was essential for development, and that the World Bank must re-engage across the board. As described by the Chinese Executive Director, the debate had not just been about dams and infrastructure, but had changed the de facto governance of the World Bank (Sebastian Mallaby, 2004). This was a historic turning point, in which increasingly-independent, powerful and confident developing countries were no longer willing to be objects of recipes decided by rich country NGOs and governments who did not have to live with the consequence of their decisions.

This re-thinking of the role of infrastructure in development—of which the current book constitutes an important part—and the re-calibration of governance structures in major multilateral institutions has had a major practical impact. Over the last five years infrastructure in general and water infrastructure in particular, has again become a legitimate area for engagement by the World Bank (and the regional development banks). Just as Jim Wolfensohn's early decision to drop Arun was highly symbolic, so too was his declaration, mid-way through his presidency that "it is important that we have a balance between the Berkeley mafia and the Chadians...and I, for my part, am more interested in the Chadians" (Sebastian Mallaby, 2004). Fittingly the focus of Wolfensohn's last Board meeting was the approval of another controversial dam—the Nam Theun 2 project in Laos. The two following Presidents of the World Bank, Paul Wolfowitz and Robert Zoellick, have both strongly supported Bank engagement in major water and other infrastructure.

References

Bell, C. and S. Devarajan (1985). Social Cost-Benefit Analysis in a Semi-Input-Output Framework: An Application to the Muda Irrigation Project, in G.Pyatt and J.I. Round eds., Social Accounting Matrices: A Basis for Planning, Washington, D.C: The World Bank.

Gittinger, J. Price (1982). The Economic Analysis of Agricultural Projects. Washington, DC: The World Bank.

Grey, D. and C. Sadoff (2007). Sink or Swim: Water Security for Growth and Development. Water Policy 9, 2007, pp.514-527.

Hazell, Peter B.R. and C. Ramasamy (1991). The Green Revolution Reconsidered: The Impact of High-Yielding Rice Varieties in South India, Baltimore, MD: The Johns Hopkins University Press.

Kapur, Devesh (2002). Do as I say, not as I do. Washington DC: Center for Global Development Working Paper 16, 41 pages, October.

McCully, Patrick (2005). The use of a trilateral network: An activist's perspective on the Formation of the World Commission on Dams, American University International Law Review, 16, 1453-1475, 2005.

Ortolano, L and K. K. Cushin (2003). Grand Coulee Dam and the Columbia River Basin. World Commission on Dams.

Sebastian Mallaby (2004). Back to the Future (Chapter 13, pp 336-373), in The World's Banker, Penguin Press.

UNIDO (1972). Guidelines for Project Evaluation. New York: United Nations Industrial Development Organisations.

U.S.Congress (1955). Discussion of Budget Bureau Circular A-47 and the Related Power Partnership Principle, Hearings before the Committee of Interior and Insular Affairs, United States Congress, March 15 and 30, 96 pages.

World Bank (2003). *Water Resources Sector Strategy, Strategic Directions for World Bank Engagement*. Washington DC.

World Bank (2002). External views on the World Bank Group's Draft Water Resources Sector Strategy: "How they were elicited, what they are, and how they will be addressed", August 2002, 24 pages.

Index

A

B

C

H

I

K

L

M

N

Q

R

S

T

U

V

W

Y